CONTENT FOUNDATIONS

Scale Your Content with AI
Without Losing Your Voice

BY ERIKA HEALD

CONTENT
FOUNDATIONS

Scale Your Content with AI Without Losing Your Voice

ERIKA HEALD

Dedicated to Lewis, for thinking I could.

ISBN: 979-8-9932806-0-8

TILT
PUBLISHING

Tilt Publishing
700 Park Offices Drive, Suite 250
Research Triangle, NC 27709

CONTENTS

Why Now Is the Time to Build Your Content Foundations

There's a universal challenge keeping content creators—from scrappy content entrepreneurs to seasoned enterprise marketers—from scaling their efforts effectively.

Spoiler alert: it's not AI.

The real roadblock? The lack of a strong content foundation.

Imagine hiring a construction crew to build your dream home—but instead of giving them blueprints, you hand them a napkin sketch or talk through your vision over a rushed coffee. Sounds absurd, right? And yet this is exactly how many organizations attempt to scale their content efforts.

We expect freelancers, agencies, and our employees to generate content that's on-brand and impactful—often without a content strategy, without a brand voice guide, and without clear processes. Many of us give our baristas more precise instructions than we give our content creators!

It's time we brought the same care, planning, and precision to our content operations that we'd bring to building our dream home. This book will help you do just that.

You'll learn how to:

- **Build a content framework** (also known as content governance) that supports scale without sacrificing creativity.
- **Document your content strategy** so it's actionable for humans and AI.
- **Create the processes, templates, and guidelines** that free your team from guesswork—and from creating random acts of content.

I'll walk you through proven frameworks I've used with Fortune 500s, high-growth startups, content-focused small and medium businesses (SMBs), scrappy nonprofit teams, and my own business. Whether you're a team of one running a content business or leading a large content team, you'll find practical, scalable steps to take control of your content ecosystem.

What Is Content Governance, and Why Do You Need It?

No matter what you call it, content governance is the only thing standing between a well-crafted strategy and random acts of content[1].

Let's be real: most content creators don't suffer from a lack of ideas. They suffer from a lack of goal alignment, consistency, and clarity. Writers are operating in silos, freelancers are guessing at brand voice, AI tools are expected to "just figure it out," and no one's quite sure who owns what, when, or why.

Does that chaos sound familiar? If so, it's costing you time, trust, and results. And it's squandering your limited resources.

---

1   "No more random acts of content" has been my primary tagline and mission since starting my consultancy. Random acts of content are all those pieces of content that are created because someone tells you to create them, despite not being tied to a documented content strategy that maps to your business goals.

Governance isn't red tape. It's your operational superpower.

It's the system that transforms your content from "just getting it out the door" to delivering real business value—consistently. With content governance in place, you gain:

- **Clarity**. Everyone—from your new intern to your seasoned agency partner—knows what good looks like and how to deliver it.

- **Consistency**. Your content actually sounds like you, no matter the format, channel, or creator.

- **Confidence**. Your team spends less time second-guessing and more time creating. Review cycles get faster, and your content gets better.

- **Scalability**. You can take on more projects, channels, and contributors—without sacrificing quality.

- **Strategic Alignment**. Content becomes a true business asset, not just a marketing expense. You're not just producing—you're performing.

And perhaps most importantly in today's content landscape: Governance is what makes AI work for you—not against you.

When your brand voice, content taxonomy, templates, and guidelines are clearly documented, AI tools stop creating generic filler and start generating real value. You get to scale your content creation without diluting your brand or overwhelming your team.

Thoughtful governance empowers humans and AI to collaborate—and create—with clarity.

This isn't about perfection. It's about creating a sustainable, strategic system for content that supports your goals today and evolves with you tomorrow.

Why Is This the Content Governance Era?

Because in our rush to scale with AI, we've flooded the internet with mediocrity. And, as Joe Pulizzi noted in a recent blog post[2], "The world does not need more noise."

While AI can accelerate content creation, too often it's treated like a magic wand for creating content out of thin air instead of the creative collaborator it's meant to be.

Let's be clear: AI is only as good as the inputs you give it.

It needs the same level of coaching and context you'd give a new hire, including a playbook that defines what good content looks like.

What's in It for You?

Content is your brand's handshake, elevator pitch, and customer experience—all rolled into one.

Content governance ensures that handshake is firm, that pitch is compelling, and that experience is consistent, every time and every place your brand shows up.

If you want to:

- Scale content without losing quality
- Empower your team (and AI) to deliver confidently
- Future-proof your marketing operations

...then content governance isn't a nice-to-have. It's about to become your competitive advantage.

Who This Book Can Help

Whether you're running a content-first business, managing a growing marketing team, or leading content strategy across a complex enterprise, this book is for you.

You might be:

---

2   Pulizzi, Joe. "The 9 Big Lessons Every Creator Needs Right Now." Joe Pulizzi, 29 August 2025, https://www.joepulizzi.com/news/the-9-big-lessons-every-creator-needs-right-now/. Accessed 25 September 2025.

- A **content entrepreneur** juggling multiple roles while trying to build a brand that feels bigger than your bandwidth.
- An **SMB marketing leader** working with a tiny team—or no team at all—to punch above your weight with consistent, compelling content.
- A **content strategist or head of content** at a large company, coordinating content operations across functions, teams, and tools while chasing alignment and accountability.

No matter your title or team size, you share the same goal:

To create content that builds trust, deepens relationships, and drives measurable results.

But right now, you're spending too much time fixing avoidable problems:

- Rewriting off-brand content
- Repeating the same creative briefs
- Scrambling to explain "what good looks like"
- Wondering if your AI tools are actually helping or just adding noise

What if, instead, you had:

- A content system that scales with you, not against you?
- Templates and workflows that shorten production time without sacrificing quality?
- AI collaborators that actually stay on-brand because they're trained on your style, voice, and values?
- A confident, creative team that knows exactly how to execute your vision—without micromanagement?

That's the sparkling future (in the words of my friend and interactive content expert Maureen Jann) content governance unlocks.

This isn't about adding bureaucracy. It's about removing friction. So you can spend less time fixing broken processes—and more time creating content that matters.

Let's get started.

P.S. You can read this book cover to cover or jump to the chapter you need—intentional repetition throughout reinforces what matters.

PART I

Building Robust Documentation

A content creator's work is never done—especially if every blog post, webinar script, or social media post needs to go through you for edits, approvals, or last-minute rewrites to "make it sound right."

The 15th edition of the Content Marketing Institute and Marketing-Profs B2B content marketing research[3] brought this into sharp focus, finding that almost half of marketers (45%) lack a scalable model for content creation.

So it's no surprise that more content leaders are turning to AI tools to scale content production. It's tempting—AI can speed up your process and take some of the repetitive, tactical work off your plate. But here's the hard truth: more and faster isn't good enough.

Speed without clarity just accelerates the chaos. Volume without alignment just creates a bigger mess. Creation without human experience just makes people scroll past.

My LinkedIn feed—filthy with more broetry[4] than you'd find at a venture capital (VC) fast pitch event—is a great example of rampant AI content creation run amok.

---

3   Stahl, Stephanie. "B2B Content Marketing: 2025 Benchmarks & Trends." Content Marketing Institute, October 9, 2024. https://contentmarketingin-stitute.com/b2b-research/b2b-content-marketing-trends-research-2025/.

4   Broetry is a LinkedIn-born post style marked by frequent line breaks, clickbait headlines, self-flattering anecdotes, and cliché advice. Its pseudo-poetic format grabs attention by seeming profound while offering little real depth. This definition courtesy of my custom GPT, which was a lot nicer in its definition than I was tempted to be.

Yes, AI can assist with content creation. But it can't replace the foundational thinking that great content is built on—your experience, your voice, your strategic lens.

You can't scale something that hasn't been defined. And you certainly can't delegate (to a human or to a technology platform) what you haven't yet documented. That's why the first step in scaling content isn't automation. It's articulation.

The Foundation for Scalable Content Starts with Robust Documentation

If you want your content to be:

- Consistent across teams, creators, and platforms
- Instantly recognizable as your brand
- Shareable, easy-to-repurpose, and strategic

...you need more than just ideas. You need systems.

In this section, we'll walk through how to create those systems—how to take the brand voice that lives in your head (and maybe in a few scattered documents) and turn it into usable, living resources that support your team and tools.

We'll cover the four pillars of foundational content documentation:

- Defining your brand voice so you can stop rewriting everything yourself.
- Creating a compelling style guide that turns content guidelines into enablement—not a PDF no one reads.
- Documenting audience personas so your creators know who they're talking to and what those audiences need from you.
- Building an effective taxonomy to organize your information in a structure that makes it easy to find and understand for both humans and robots.

By the end of this section, you'll have the tools to move from reactive content production to proactive, strategic creation—supported by documentation that works with you, not against you.

CHAPTER 1

Defining Your Brand Voice

Your brand voice is where content as a differentiator and business-driver begins.

Your brand voice isn't just how you sound—it's how you make people feel. It's how you show up in inboxes, on landing pages, and in LinkedIn comment threads. It's what makes your content recognizable, trustworthy, and worth paying attention to—even in a crowded, noisy feed.

And if you haven't defined your voice? You're leaving that brand experience to chance.

Why Brand Voice Matters Now More Than Ever

I don't have to tell you that we are currently slogging through an environment of content uniformity and AI-generated slop. The good news is, this is why a clear, differentiated voice is one of your brand's biggest competitive advantages. It's also the foundation that every human and AI content creator needs in order to produce content that aligns with your strategy and reflects your values.

You wouldn't hand a designer a blank canvas and say, "Just make it look like us."

So why do that to your content creators?

Defining your brand voice gives your team (and your tools) a shared creative starting point. It removes the guesswork. It sets the bar. And it enables you to scale content without sounding like you've outsourced your personality.

In this chapter, I'll walk you through the step-by-step process of defining and documenting a usable brand voice. This isn't a theoretical exercise—it's a practical toolkit you can use with your team, your stakeholders, and even your AI tools.

You'll learn how to:

- Identify the core traits of your brand's personality
- Translate those traits into practical dos and don'ts for content creators
- Create a brand voice chart that's easy to share, apply, and evolve
- Avoid common brand voice traps (like sounding "professional" but bland)

You'll also see real-world examples of how brand voice charts are used in organizations of all sizes—from scrappy startups to global teams.

When you define your voice, you create a compass for every piece of content. You make it easier to collaborate.

You make your brand feel more human. And you give your content the clarity, confidence, and consistency it needs to stand out—and scale.

The Brand Voice Difference

Most brands don't have a brand voice problem—they have a brand voice documentation problem. Everyone says they want to sound "human," "authentic," "insightful," or "bold." But when it's time to write a blog post or brief an AI tool, what does "authentic" actually mean? Does that tone include emojis? Humor? Data-backed claims? All three?

Without clear documentation, even the most well-intentioned content creators end up guessing—or worse, creating off-brand content that needs to be rewritten or walked back.

Your Brand Voice Is Your Content Superpower

Your brand voice is who your brand is and how that identity shows up in the content you create and the way you communicate.

It's what makes your brand feel familiar across channels, creators, and platforms. It's what builds trust with your audience—even before they know what you sell. And it's what enables you to scale your content without losing your soul.

But this only happens if you define and document it for how it shows up in the content you create.

Why You Can't Skip This Step

If you've ever:

- Rewritten a freelancer's draft because it "just didn't sound like us"
- Had a company leader strike through every attempt at humor or boldness they originally asked for
- Wondered why your AI-generated content sounds like a robot or your competitor

...then you've experienced the cost of a fuzzy or poorly defined brand voice. Your team can't follow guidelines you haven't given them. Your AI tools can't replicate what hasn't been clearly articulated.

This is the essential work that saves you from the endless cycle of content rework, internal debates, and random acts of content.

What If You Don't Have a Well-Defined Brand Yet?

That's okay. Start with what you do have. If your brand guidelines include adjectives like "trusted" or "playful," plug those into a prompt and ask AI to help you flesh out definitions and examples.

Not sure where to begin? Ask AI to review your website or blog content and identify patterns. This will give you a "minimum viable brand voice" you can refine over time. Or consider this approach branding consultant Kate DiLeo shared in her #ContentChat[5] episode[6]: "I love to

---

5 #ContentChat is the weekly LinkedIn livestream conversation I lead on Mondays at noon Pacific/3 p.m. Eastern. Each week I explore a content marketing or content governance topic with a practitioner, author, or consultant who is an expert on that week's topic. Over the past decade leading the chat, I've spoken with hundreds of experts representing all sides of content, marketing, and brand building—and you'll see many of them featured in the following chapters. See https://erikaheald.com/content-chat/ for more details and episode recaps.

6 DiLeo, Kate. "#ContentChat Essential Branding Elements for Successful Content Marketing." Erika Heald Marketing Consulting, April 5, 2024. https://erikaheald.com/content-chat-branding-considerations-for-content-marketing/. Accessed September 29, 2025.

take a very simple approach to helping an organization or individual really understand their brand's tone of voice or personality," she said.

"The question that I love to ask is a fun one: If your brand were a person, who would it be? I love to push it further—not just talking about the personality traits, but I want them to name a public figure or celebrity. And I want them to think about how that person sounds. When it comes to a brand's tone of voice, we're trying to get our arms around not just personality—oh, they're cool, they're nice—but what's the rhythm? If you were to close your eyes and just listen, could you know what brand is talking to you?"

Crafting Executive Communications Without Brand Voice Guidelines

If you're still not convinced about the importance of a well-defined brand voice, let me share with you what it's like to be a new hire trying to create on-brand content without one.

On my second day of work at a long-ago job, my new boss asked me to write an all-employee email from the head of our division, congratulating our employees on having just completed another successful client conference. I hadn't attended the event, and hadn't yet met the executive I'd be ghostwriting for.

So I did what any other writer would do—I asked for the brand voice guidelines and style guide to ensure I got their voice right. My new boss stared at me, and not in the "great question!" sort of way.

Sadly, this was not the last time I'd be told there weren't any content guidelines in place, or that an executive I'd be asked to write for didn't have a defined voice or thought leadership platform.

I followed up by interviewing a colleague about the event, reading an OpEd with the executive's byline, and drafted the email copy. It came back with more red pen than black laser print copy. It wasn't a shock that I wasn't able to get it right with so little input, but it also wasn't a good use of my time. And it didn't help me feel like I was getting off to a good start at that job, either.

Eventually, I got the executive's voice right. I even wrote one of their most well-received and responded to emails of all time. But it remained a struggle because those foundational elements for creating consistent content were never put into place.

You see, my boss had made themselves indispensable by being the one person every piece of content and every communication activity had to go through. And then the company had layoffs, and my boss was no longer there to act in that role. This left me and the rest of our team trying to figure out what the processes were to do our work. It was like we were starting from scratch to figure out how to do our jobs.

Within less than a year, everyone on my team had moved into other roles—some within the company, and others with competitors. I can't claim that a documented brand voice could have saved the team from all that turnover. But I do know it would have made a significant difference in the work I did there and been a great onboarding tool for the inevitable next crop of writers.

How to Create a Brand Voice Chart

Now that you can see the power of not having a well-defined brand voice, let's talk about how to do it well.

I've found that the best way to define your voice is through a Brand Voice Chart. It's a one-page cheat sheet that breaks your voice down into traits, with clear guidance on how those traits do and don't show up in your content.

Here's an excerpt from my brand voice chart:

Brand Voice Attribute	Attribute Definition	Dos	Don'ts
Expert	Demonstrates and communicates expertise in the content marketing industry, showcasing professionalism.	• Use industry terminologies • Back up statements with facts or experiences • Cite your sources • Share practical advice based on experience	• Make unfounded claims • Use unnecessary jargon • Include acronyms without first defining them • Reference old, third-party sources
Ethical	Maintains honesty, transparency, and ethical standards in all communications.	• Be transparent • Own your mistakes, apologize, and correct them promptly • Follow (and set!) industry ethical guidelines	• Exaggerate • Make things up • Promise results outside of your control • Recommend or approve unethical tactics
Empowering	Strives to equip marketers with the knowledge, tools, and guidance that they need to succeed with content.	• Provide concrete, practical tips • Create and share valuable frameworks, templates, and other how-to resources	• Focus only on the what and why • Create content that the consumer can't immediately use or learn from • Create content to fill a calendar slot
Curious	Cultivates an interest in learning, exploration, and staying updated in content marketing.	• Continuously share new learnings • Track emerging trends • Comment on industry hot topics • Prioritize exploring and learning	• Set it and forget it • Be resistant to change • Ignore industry innovations, advancements, and evolution

1. Step 1: Define What Each Trait Means—Specifically

This is the secret sauce. Don't just list "irreverent" or "helpful" and call it a day. For each trait, write a short description of what that word means in your context. (Your "irreverent" may be someone else's "inappropriate.")

Here are a few additional definition examples:

- **Bold** – Uses strong, decisive language and clear points of view to stand out and challenge conventions.
- **Practical** – Focuses on clear, useful, and immediately applicable advice or solutions.
- **Playful** – Brings energy and lightness through witty phrasing, creative analogies, and a sense of fun.

See how these definitions already are helping you see how you'd approach creating content that reflects those traits? And, more importantly, what approaches wouldn't work within those definitions? That's the power of a well-defined brand voice trait.

Step 2: Add Real-World "Dos" and "Don'ts"

For each trait, list examples of how it shows up in your content—and how it shouldn't.

The Dos should include practical examples of word choice, tone, and techniques that reinforce the trait (e.g., asking open-ended questions to show curiosity, or using straightforward language to demonstrate practicality).

The Don'ts highlight specific missteps that undermine the trait, such as not citing data sources despite striving for an expert voice or overusing irrelevant jokes that distract from content clarity when attempting to be playful.

By providing both positive and negative examples, you equip content creators with clear guardrails to consistently apply the brand voice and avoid common pitfalls.

This becomes your creative team's filter. When they ask, "Does this sound like us?" they can look to your voice chart for a gut check.

Step 3: Package It as a Shareable Chart

Make your chart visual, accessible, and easy to reference. Drop it into your content brief templates. Link to it from your editorial calendar. Embed it in your AI tools.

Make your brand voice chart as common and unavoidable as the nutrition label on packaged food!

If people can't find it, they can't follow it.

Once It's Documented, Use It for Everything, Everywhere

Your brand voice chart isn't just for the content team or your agency. It should be used throughout your company to inform:

- Your social media tone
- Your product copy
- Your customer service scripts
- Your AI-generated outputs

If your chatbot sounds robotic or your email campaigns feel inconsistent, the problem often starts with voice.

The Payoff: Scalable, Differentiated, Human Content

When your brand voice is clearly defined and documented, it becomes a powerful guide for everyone involved in creating and approving content.

Writers work faster and with fewer edits because they know exactly how the brand should sound. AI tools can generate content that feels authentically like you, rather than generic filler. Stakeholders align more easily around what "on-brand" really means, reducing conflicts and confusion. And most importantly, your audience comes to recognize and trust your content—no matter where they encounter it.

Defining your voice is the first and most important step in building a content program that scales. As my friend and content strategist Carmen Hill said to me, "AI can help surface the patterns—but it's still on us to make sure our brand voice is different, consistent, and human."

Brand Voice Pitfalls

A well-crafted brand voice is essential for building trust, standing out from competitors, and giving internal and external creators what they need to deliver content that resonates. But even the most well-intentioned brand voice strategies can fall short if you're not aware of the common pitfalls that sabotage their success. Let's explore the most frequent missteps—along with strategies to avoid them.

1. Overly Generic Voice Traits

Using vague descriptors like "friendly," "professional," or "innovative" without clear definitions leads to wildly inconsistent interpretations. For example, calling your voice "irreverent" could mean "cheeky and bold" to one writer and "snarky and borderline offensive" to another.

Fix it: Use a brand voice chart that includes not just the voice trait but also a clear definition, dos and don'ts, and content examples that embody the trait in practice. Creating a brand voice isn't just choosing a few adjectives. It's defining what those words look like in action—so creators stop guessing.

2. Aspirational, but Unachievable Voice Goals

Aspirations are fine—but only if they're achievable with your current resources, skill sets, and organizational permissions. Teams that list "authoritative" as a brand trait often fall into this trap. Without original research, expert voices, or deep industry content, you're setting yourself up to under-deliver.

Fix it: Ground your voice in what your team can consistently execute. If you aim to be authoritative, ask: Do we have the capacity to regularly produce thought leadership content that supports this claim? If not, consider traits like "educational" or "helpful" as more realistic and sustainable alternatives.

Aspirational voice goals only work if your team has the content muscle to back them up.

3. Sounding Like Everyone Else

If your brand voice is indistinguishable from your competitors, you risk becoming forgettable. When tone, visuals, and messaging all converge on the same "approachable luxury" middle ground—as with many direct-to-consumer brands—it's hard for customers to tell you apart. What you should be striving for is crafting a brand voice that your fans could identify without your logo attached.

Fix it: Conduct a competitor brand voice audit. Identify overlaps in tone, language, and message, and then intentionally push your voice in a direction that highlights what makes your brand meaningfully different.

If you're not auditing your competitors' content voice, you're missing the opportunity to differentiate.

4. Unclear Guidance for Content Creators

A documented voice that lacks context or isn't integrated into content workflows leads to off-brand content. If content creators aren't sure how your brand "shows up" across touchpoints—or they're relying on guesswork or old campaigns as reference—they'll inevitably use inconsistent language and tone.

Fix it: Incorporate your brand voice chart and style guide into content briefs, templates, and onboarding materials. Make your expectations clear and accessible. A well-maintained voice chart acts like a GPS: It helps internal and external creators navigate your tone without constant course correction.

Your team can't follow guidelines you haven't given them and your AI tools can't replicate what you haven't clearly articulated.

5. Failing to Evolve

A voice defined years ago can become a constraint if it no longer reflects your brand's direction or your audience's expectations. Static voice guidelines stifle growth—and worse, they may encourage content that feels outdated or irrelevant.

Fix it: Schedule regular brand voice audits. Use performance data, customer feedback, and AI tools to assess how your voice is landing. Then iterate. Voice is a living system—it should evolve with your brand, not fossilize behind it.

Why AI Is the BFF You Didn't Know Your Brand Voice Needed

As a writer, people are often surprised when I tell them I really do think AI can be a creator's best friend. Hear me out!

It's only natural that marketers—stewards of their brand's voice—may feel uneasy about letting artificial intelligence into their office. You may even feel the urge to say "you can't sit with us".

But AI tools aren't mean girls. It's equally possible that we can harness AI and collaborate with it to make it our brand voice ally, helping us document and apply a standardized voice across our entire organization. AI and your brand voice may come from different cliques, but when they team up, they can create magic.

Imagine a world where every communication your organization sends out sounds like it could come only from your brand or company. Where everyone uses the appropriate tone, in every situation, to match your customer's mindset.

That's the vision that leads me to propose that it's time marketers replace our AI-as-nemesis expectation with a more friendly and beneficial one: Let's treat AI as a new Best Friend Forever (BFF) that can help us maintain consistency in our brand voice across all content.

When you're on a content creation mission, maintaining a consistent brand voice can seem like battling the Hydra—you overcome one inconsistency, and two more spring up.

Ugh.

With the massive amount of content getting created daily, nailing down your brand's voice across every document, web page, and social media post can test your mettle.

This is where AI swoops in as your BFF, extending its helping hand to ensure consistency in your brand voice. Think of it as a diligent comrade, tirelessly coding in the background, learning your brand's particular language and style nuances while you sleep!

How to Enlist AI for a More Consistent Brand Voice

Now that we're besties with AI, let's walk through a few simple ways that AI can help support your brand voice.

1. AI can identify your competitors' brand voices

The first step toward creating and nurturing a differentiated brand voice is to understand how your competitors talk to their people so you can differentiate. The old way we used to do this was incredibly time intensive. We'd go to conferences, pick up pieces of collateral, and talk to colleagues who used to work for our competitors. Today, by using

AI, you don't need to guesstimate. You can actually look at competitors' content across their channels and come up with something a lot more specific (and accurate) in seconds.

Here's a prompt to try for yourself:

> You are an experienced content marketer, with a specialty in branding and creating a consistent brand voice, and you work for [company type] in [industry]. You need to define and document your competitor's brand voice and how it is reflected in their content.
>
> Your competitor is [competitor name], the maker of [product name], [product name], and [product name]. See [website URL] for more details.
>
> Define 5 key brand voice attributes for [competitor name] and display them in a simple table with the following columns: Voice Attribute, Attribute Definition, Dos, Don'ts. Create a separate row for each attribute.

This should give you a result that looks something like the brand voice chart I shared with you earlier in this chapter.

Once you have voice charts in hand for each of your competitors, you can reflect on your own brand voice and what makes it different from your competitors'.

2. Let AI audit how well your content reflects your ideal brand voice

Staying on message is hard for writers—and even more difficult for the many other people in your organization who create content and communications. With AI, you can get an objective assessment of how good a job an individual piece of content—or even your entire website—is doing at staying on voice.

Now, go ahead and use the competitor voice prompt with your own company's details. Does the output AI created come close to reflecting your brand? Or does it feel pretty indistinguishable from your competitors? If it isn't quite what you expected, ask AI for help. Here's a prompt:

> You are an experienced content marketer with a specialty in branding and creating a consistent brand voice. You work for [company type] in [industry] and are the maker of [product name], [product name], and [product name]. See [website URL] for more details.

You need to refine how your brand voice is reflected in your content. Your 5 key brand voice attributes are [attribute 1], [attribute 2], [attribute 3], [attribute 4], and [attribute 5]. What are your recommendations for how to update the content on your website to better reflect those brand attributes?

Let's look at an example of the kind of feedback you can expect to receive from this prompt. For the record, I asked ChatGPT to give a legacy software company a bit of a youthful brand voice makeover. Here's what it proposed to update the website's tone and language:

Fun and Youthful:
- Use playful and engaging language
- Incorporate casual and conversational tone
- Add humor and light-hearted elements where appropriate

Creative and Whimsical:
- Employ imaginative metaphors and analogies
- Use vivid, expressive descriptions
- Integrate storytelling to make content more engaging

Unconventional:
- Challenge norms with unexpected headlines and phrases
- Make bold and surprising statements
- Break out from traditional content formats

Once you have this feedback in hand, you can even ask for specific voice makeover examples.

For this example, I took a very traditional headline from Microsoft Office® and asked to have it updated to reflect the above extreme makeover. Here's what happened:

Original Headline:

"Enhance Your Productivity with Microsoft Office"

Revised Fun Headline:

"Boost Your Productivity Mojo with Microsoft Office Magic!"

AI's Explanation:

- "Boost Your Productivity Mojo" adds a playful and energetic vibe, making it feel more engaging and fun.

- "Microsoft Office Magic!" introduces a whimsical element, suggesting that using Microsoft Office can feel almost magical in how it helps with productivity.

This revised headline reflects the fun, creative, and whimsical attributes while still conveying the core message about enhancing productivity with Microsoft Office.

No more guessing at how to reflect those brand voice attributes—AI can be a pocket voice coach to help you get there in a click.

3. Have AI create the rough drafts for derivative content

Picture this: You've spent weeks perfecting a video script for your campaign's launch or crafting an in-depth e-book. Now you have to create a series of blog posts, social media snippets, and infographics derived from this cornerstone content. Instead of hand-wringing over how to maintain the same tone and message, why not let AI do the trick? It can generate rough drafts based on your templates, guidelines, and, crucially, your brand voice.

For this kind of task, I like to use Copy.ai or another paid AI tool that allows you to upload or copy+paste your content style guide and voice documentation into your private workspace. This saves significant time in crafting prompts.

If a paid tool with a private info-base isn't an option, you can still use AI, but you will need longer, more detailed prompts, and may need to go through a few prompts to get to your desired content end point.

Here's a prompt to try for creating social media posts to promote a piece of content:

> You are a content marketer who is skilled at crafting [derivative content type] that [content goal] from [content audience]. You need to promote this [content type] [content's URL] on [channel]. Please draft 4 [content type] that reflect [company name]'s brand voice and style, as seen at [main company URL].

I ran this prompt using a blog post on my site. Here's what I got back from my custom GPT that has my brand voice and style guide embedded in it:

Here are four LinkedIn posts crafted to promote the blog post from Erika Heald, reflecting her brand voice and style:

Post 1:

> 🗣️ Struggling to define your brand voice? It's time to get clear and consistent! Our latest blog post dives deep into how you can articulate your unique brand voice to drive your content marketing strategy. Don't miss out on these actionable tips and expert insights! 📢 [Link to Blog Post]

Post 2:

> 💓 Your brand voice is the heartbeat of your content. Are you giving it the attention it deserves? Learn how to define and refine your brand voice with practical advice from our newest blog post. Let's transform your content marketing approach today! 🚀 [Link to Blog Post]

Post 3:

> 💬 "Consistency is key." We hear it all the time, but how do you apply it to your brand voice? Our latest blog post breaks down the steps to creating a cohesive and compelling brand voice that resonates with your audience. Check it out now! 📖 [Link to Blog Post]

Post 4:

> 🔍 What makes your brand stand out in the crowded digital space? A well-defined brand voice! Discover the secrets to crafting a distinctive and memorable brand voice in our latest blog post. Elevate your content marketing game with these essential tips! 🎯 [Link to Blog Post]

> Each post is designed to be engaging, informative, and reflective of Erika Heald's brand voice, encouraging B2B marketers to explore the blog post and apply its insights.

While I wouldn't copy+paste this AI-generated content as is (especially not the emojis), these drafts are solid starting points.

In each of the above use cases, AI gives you back valuable time by acting as an intern and research assistant, following your documentation and delivering on your requests so you can spend more time creating and refining your content. Just what you'd expect from your new BFF.

AI Isn't Replacing the Human Touch; It's Supporting It

It's crucial to remember that AI will not replace the human touch in content. After all, it's your creativity, passion, and insights that form the heart of your brand voice. But, just like a true BFF, AI can offer support, enhance your work, and lighten the load when you're coming up on a tight deadline.

As marketers, we can collaborate with AI, harnessing its capabilities for the good of our brand voice. When done correctly, AI becomes a tool through which we can imagine a more cohesive, effective, and resonant brand. One that speaks to our audience in a voice they recognize and trust.

Creating and Maintaining a Living Brand Voice

Earlier in this chapter, we explored why brand voice is a cornerstone of differentiation and how to define the personality traits that make your brand unique. Now it's time to bring that work together—moving from defining your voice to documenting it in a way your entire organization can use. This is the moment where your brand voice stops being an idea and becomes a practical, everyday tool.

1. Audit Where You Are

Whether you're a solo marketer or part of a global content team, the first step is to take stock of your existing content. Review a representative mix across channels—blog posts, social updates, videos, email campaigns, e-books, webinars. Identify the pieces that most successfully met your goals and felt unmistakably "you."

Just as importantly, set aside the pieces that could have come from a competitor or that no longer reflect your current voice. This isn't about critiquing past work—it's about establishing your baseline so you know what needs to evolve.

2. Share Your Brand Voice Chart Far and Wide

A brand voice chart is only useful if it's used. Don't let it sit in a forgotten folder. Share it with anyone who creates or approves content: marketing, sales, customer service, product—even external partners.

Pair it with real examples from your content library: a "hall of fame" piece that nails all your traits, a single-trait example that shows one element in action, and a "close but needs tweaking" example to illustrate fine-tuning.

Most importantly, embed the chart into your processes. Add it to your content brief template. Include it in your AI tool reference libraries. Make it part of onboarding for new hires and agencies. The more accessible and integrated it is, the more consistently it will be used.

3. Keep It Alive

Your brand voice will evolve as your business grows and your market shifts. Review and refine your chart at least annually. Ask yourself: Has your audience changed? Are your traits still differentiating you from competitors? Are there traits your team consistently struggles to bring to life?

And if competitors start imitating your voice—a sure sign you're onto something—adjust to stay ahead.

Turn Your Voice Into a Scalable Asset

Defining your traits was the starting point. Documenting them in a brand voice chart—and making that chart a living part of your content process—is what ensures those traits consistently show up across every channel, from the CEO's keynote to a quick social post.

This is how brand voice moves from an abstract idea to a scalable asset that protects your differentiation—whether your content is created by a seasoned marketer, a new employee, or an AI tool.

CHAPTER 2

Assembling a Comprehensive Style Guide

Imagine handing off a blog post, social media campaign, or even a customer case study to a teammate—or to an AI assistant—and getting it back exactly how you envisioned it. No rewrites. No endless Slack clarifications. No awkward conversations about tone or formatting.

That's the power of a strong content style guide.

A style guide is more than a grammar cheat sheet or a logo usage reference. It's your brand's DNA in written form. It's how you keep content cohesive and aligned with your voice, no matter who's writing it—whether that's a new hire, a freelancer, a cross-functional partner, or a generative AI tool.

This chapter is your blueprint for building a style guide that actually gets used—and makes your life easier. You'll learn:

- Why a content style guide is essential to producing consistent, high-quality content at scale.
- The foundational elements every useful guide should include.
- How to ensure your guide grows with your team and your content needs.
- A starter template to help you create your own guide—without starting from a blank page.

If you've ever felt like you're the only one who really gets your brand's voice, tone, or content expectations, this chapter is for you. Let's document what's been stuck in your head—and make your brand voice repeatable, scalable, and unmistakably yours.

Why a Content Style Guide Is Essential

If content is the embodiment of your organization's voice, a style guide is your vocal coach.

When you're a team of one—or even just a tight-knit, well-aligned group—it can feel like a style guide is a "nice to have." After all, you know how things should sound, what to capitalize, and when to use the Oxford comma. But as your content team grows (or as you rely more on freelancers, subject matter experts [SMEs], or AI tools), your ability to scale high-quality, on-brand content depends on one thing: shared clarity.

A content style guide removes the guesswork. Without it, every writer has a slightly different take on your brand voice. AI tools generate inconsistent content. Editors spend time fixing tone, not improving ideas.

Freelancers get stuck waiting for feedback—or worse, keep repeating the same mistakes. And you become the bottleneck for every single asset.

A style guide brings consistency and efficiency to your workflows by serving as a single source of truth for voice, tone, formatting, grammar preferences, and brand-specific language.

With these expectations documented, internal and external creators can self-correct before submitting content and AI tools perform better because they're trained against clear standards. Review cycles become shorter and more productive, since less time is spent debating style choices or fixing preventable errors.

The result is content that feels cohesive across channels, creators, and campaigns—and you get back valuable time you can use to focus your attention on the work that truly requires your expertise.

Think of a style guide as your content recipe book.

Your style guide is the difference between running a restaurant with

a tested recipe book vs. telling each new chef to just cook the dishes on the menu however they want to. Even the best chefs need a baseline—and your brand does, too.

And it's not just about consistency. A great style guide actually empowers creativity. When your content creators understand the boundaries and brand expectations, they can take bigger risks within the lines—rather than wasting energy second-guessing basic decisions.

18 Essential Style Guide Elements

A well-crafted style guide doesn't just enforce consistency—it empowers creativity, improves content quality, and helps your team scale without losing your voice. That's why it's a must-have tool for onboarding, alignment, and quality control across your content ecosystem. Whether you're creating blog posts, email campaigns, landing pages, or scaling content with AI, your guide needs to anticipate what creators and collaborators actually need to do great work.

Your style guide should include more than grammar and punctuation rules. Add naming conventions, editorial preferences, tone guidance, and examples to give people the full picture.

Here are the 18 essential elements[7] Alek Irvin and I defined—in collaboration with #ContentChat community members Tod Cordill, Melanie Graham, and Carmen Hill—that every practical, scalable content style guide should include.

Note that you don't need to tackle all of these elements at once! Instead, consider this an overview of things to look for in your daily content work and capture it in a document, using the style guide template at the end of this chapter.

1. Audience + Key Personas

Clarify who your content is for and what they care about. This grounds voice, tone, and structure decisions in empathy and relevance.

What to Include:
- Primary audience(s) and roles
- Key challenges or goals
- Preferred content formats or channels
- Example persona descriptions

2. Brand Colors

Ensure visual consistency across every touchpoint. This keeps design cohesive and instantly recognizable—internally and externally.

---

7   18 Irvin, Alek. "Elements of a Truly Useful Brand Content Style Guide." June 2, 2022. https://erikaheald.com/18-elements-of-a-truly-useful-brand-content-style-guide/. Accessed September 29, 2025.

What to Include:

- HEX, RGB, and CMYK codes
- Pantone swatches (if applicable)
- Primary vs. secondary color use
- Color contrast or accessibility notes

3. Company Naming & Taglines

Establish how to reference your brand—formally and casually. This avoids misnaming, acronym misuse, and inconsistent messaging.

What to Include:

- Legal name and acceptable shorthand
- Tagline(s) and usage rules
- Product or service naming conventions
- Terms to avoid or deprecated names

4. Company Overview

Ground your content team in your brand's identity and positioning. This creates cohesion across internal comms, public relations (PR), and marketing.

What to Include:

- One-line company description
- Short and long-form boilerplate copy
- Founding story and key milestones
- Leadership roles or references

5. Email Signatures

Standardize how your team shows up in inboxes. This builds brand recognition, professionalism, and trust.

What to Include:

- Signature layout and required elements
- Call to action (CTA) formatting (if used)
- Legal disclaimers or compliance notes
- Links to company or social profiles

6. Font + Typography

Keep written content visually aligned with your brand aesthetic. This ensures clarity, accessibility, and design integrity.

What to Include:
- Primary and secondary typefaces
- Font sizes for headings, body, and captions
- Usage examples and download links
- Cases for exceptions (e.g., in email or print)

7. Frequently Asked Questions (FAQs)

Answer the questions creators always ask—before they ask. This reduces rework and speeds up onboarding.

What to Include:
- Common tone/style questions
- Specific voice application scenarios
- Clarifications on brand terminology
- Quick links to internal tools or guidelines

8. Images + Photography

Define your brand's visual story. This keeps content cohesive and aligned with your brand vibe.

What to Include:
- When to use photos vs. illustrations
- Preferred image styles (lighting, composition, tone)
- Approved sources or subscriptions
- Link to brand image library

9. Logo Use

Protect the integrity of your logo across all formats. This ensures consistency and prevents misuse.

What to Include:

- Logo versions (color, grayscale, reversed)
- Minimum size and clear space requirements
- Where and how to download assets
- Examples of improper usage

10.   Messaging Framework

Anchor your content in your brand's positioning. This ensures strategic alignment and consistent storytelling.

What to Include:

- Company or product positioning statements
- Messaging pillars or themes
- Key proof points and customer benefits
- Sample messaging by audience or product

11. Mission + Values

Articulate the "why" behind your brand. This connects content creation to your purpose and ethos.

What to Include:

- Mission and vision statements
- Core brand values
- How each value shows up in content
- Tone alignment with values

12. Publishing Conventions

Establish formatting rules that apply across platforms. This reduces editing time and avoids last-minute fixes.

What to Include:

- Headline and subhead casing rules
- Quote formatting and citation standards
- Metadata requirements (title, description, alt text)
- Rules for CTAs, bulleted lists, and pull quotes

13.Style Guide + Dictionary of Record

Remove ambiguity around grammar and usage. This provides clarity and consistency across all written content.

What to Include:

- Preferred editorial style guide (AP, Chicago, etc.)
- Dictionary of record (e.g., Merriam-Webster)
- Internal grammar preferences
- Formatting rules (e.g., lists, citations, punctuation)

14.   Templates

Give your team plug-and-play resources. This speeds up production and minimizes inconsistency. We'll be going through templates in detail later in the book.

What to Include:

- Content briefs and intake forms
- Social media templates
- Presentation decks and letterhead
- Links to editable versions

15.Video Specifications

Ensure your video content sounds—and looks—on-brand. This keeps multimedia assets aligned with brand personality.

What to Include:

- Intro/outro guidelines
- Music, voice, and tone preferences
- Visual style (e.g., animation, real people, typography)
- Links to examples and asset libraries

16.   Voice + Tone

Help everyone write like your brand—consistently. This helps guide creators across content types, from blogs to bots.

What to Include:

- Core voice traits with Dos and Don'ts
- Context-specific tone shifts
- AI-specific tone notes (if applicable)
- Links to exemplary content

17. What Not to Do

Prevent brand-damaging missteps. This saves editors time and ensures contributors don't repeat mistakes.

What to Include:

- Off-brand words or phrases to avoid
- Examples of what not to emulate
- Topics or competitors that are off-limits
- Reviewer "pet peeves" to note

18.   Additional Resources

Empower creators with helpful tools and references like the ones we'll walk through later in the book. This boosts content quality and reduces dependency on gatekeepers.

What to Include:

- Content calendar or brief templates
- Internal writing guides or playbooks
- External resources (e.g., Grammarly, Hemingway)
- Training materials or onboarding links

Creating a Style Guide That Grows with You

A content style guide is not a "set it and forget it" document. It's a living, evolving tool that should grow alongside your team, reflect your brand's maturity, and adapt to new content formats, platforms, and technologies (including AI).

If your guide hasn't been updated in over a year—or if it still refers to Twitter (X) as a primary customer engagement channel—it's probably time for a refresh.

Here's how to ensure your guide scales with your content efforts rather than becoming a dusty artifact in the shared drive.

1. Build for Iteration, Not Perfection

Don't wait until your style guide is "finished" to share it. Style guides evolve best when they're:

- Launched early (even as a v1.0 draft)
- Easily accessible (hosted in Notion, Google Docs, or your CMS)
- Open to feedback (with a clear point of contact for suggestions)

Make it clear this is a living document by adding a latest revision date at the top of your guide to log major updates. This also reinforces that it's a dynamic tool—not a static PDF.

2. Treat It as Part of Your Governance Infrastructure

Your style guide should integrate with your other foundational systems:

- Content briefs: Link voice/tone and formatting guidance directly inside
- Templates: Reference formatting rules, brand terms, and preferred messaging
- Workflows: Include the style guide in onboarding and review stages

  - If your team uses AI, treat your style guide as the "training manual" for prompts. The more consistently it's used, the better your content (and your AI output) will perform.

3. Invite Cross-Functional Feedback

Your content creators aren't the only people using your guide. Sales, product marketing, customer success, and external partners all benefit from clear brand guidelines.

Here are some easy ways to collect feedback:

- Add a simple form or comment-enabled doc for suggestions
- Host quarterly "style sync" check-ins to address common friction

- Monitor support tickets, Slack threads, or edit comments for recurring gaps

If a team never uses the guide, find out why. It might need to be more user-friendly—or simply easier to find.

4. Use Data to Drive Updates

Let performance and process guide your iterations—not gut feelings. Watch for:

- Content errors or inconsistencies flagged during review
- Creator or reviewer confusion (especially with freelancers or SMEs)
- AI-generated content that misses the mark
- Low-performing content formats that need fresh guidance

When these issues turn up, update your style guide accordingly with refined definitions or tone examples, clarified grammar/punctuation standards, and new templates or formatting preferences.

5. Schedule Regular Reviews

Set a cadence for maintaining and improving your guide—just like you would with your content strategy.

"Too many teams think of the style guide as a one-time task. But for it to truly support content creators, it has to be a living document," says content strategist Carmen Hill. "A style guide, like any tool, is only useful and valuable if people actually use it. This is really the biggest challenge. Regardless of how you decide to share the guide, include a link to the shared document in any channels and tools that inform content writing and creation, like your content brief, content templates, and guidelines for guest writers/contributors."

Review and update your guide at least twice a year to ensure it remains current and accurate. Underneath the "last updated" date, assign ownership to a specific person or role and provide their contact information.

During your review:

- Prune outdated references or deprecated channels
- Update brand terminology, product names, or positioning statements
- Add new elements based on emerging platforms

Keep It Usable or It Won't Be Used

Above all, keep your style guide human friendly. Use plain language and common usage, not generic academic rules. Include real examples from your own content. Break content into scannable sections with headers and bullets, and link to relevant templates, checklists, or tools. And don't underestimate the power of training.

Include your guide in:

- New employee onboarding
- AI prompt libraries
- Guest contributor packets
- Team workshops or content audits

Steal This Style Guide Template

[Your Brand Name] Content Style Guide

Last updated: [MM/DD/YYYY]

Owner: [Name + contact info]

Version: 1.0

1. Audience + Key Personas

Who we're writing for, and what they need from us

- Primary audience(s):
- Key roles or titles:
- Top challenges, metrics, and goals:
- Preferred content types or formats:
- Defined persona links:

2. Brand Voice + Tone

How we want to sound to the world

- Expand upon your voice traits
- Embed your brand voice chart

3. Grammar + Style Preferences

How we write, format, and punctuate

- Style Guide of Record: [e.g., AP / Chicago]
- Dictionary of Record: [e.g., Merriam–Webster]
- Oxford comma: [Yes / No]
- Headline 1 casing: [e.g., Title Case, Sentence case]
- Headline 2 and below casing:
- List formatting: [Bulleted or numbered]
- Emojis: [Yes / No]
- Other house rules:

4. Messaging Framework

What we say about ourselves

- Company positioning statement:
- Key differentiators or value pillars:
- Sample messaging by product or audience:
- Link to messaging doc or brand book:

5. Visual Identity References

The basics of how we look

- Logo usage: [Link to brand kit + logo rules]
- Fonts: [Primary + Secondary]
- Brand colors: [HEX + RGB codes]
- Image style guidance: [Photography vs. Illustration, tone, mood]

6. Publishing Conventions

How we prepare content for publication

- Headline + subhead casing:
- Quote/citation formatting:
- Meta title + description standards:
- CTA formatting + linking conventions:

7. Templates + Examples

What great looks like

- Blog post example
- E-book example
- Email example
- Landing page example
- Social media examples by channel

Use these templates to create in our formats

- Link to blog template
- Link to e-book template
- Link to email template
- Link to landing page template
- Link to social media guidelines and post templates

8. What Not to Do

Avoiding common mistakes

- Words/phrases to avoid:
- Deprecated product names or internal jargon:
- Brand "pet peeves" from reviewers:
- Topics or competitors that are off-limits:

9. Additional Resources

Tools to help you write well and stay on brand

- Editorial calendar
- Content request/intake form
- Internal tone training or examples
- AI prompt library or voice pack
- Grammarly/Hemingway/proofing tools

10.   Maintenance + Feedback

How to propose an edit to this guide
- Link to a suggestion form or contact email
- Next scheduled review date: [Quarterly, Biannually, etc.]
- Version history or change log: [Link or embed here]

CHAPTER 3

Sketching Meaningful Audience Personas

Before you write a single word of content, hit "record" on a podcast, or create an AI prompt, there's one critical question you need to answer: "Who is this for?"

Too often, content is created with good intentions but fuzzy targeting—trying to speak to everyone and resonating with no one. That's where personas come in. They're not just marketing exercises or customer avatars—they're the lens that brings your entire content strategy into sharp focus.

In this chapter, you'll learn how audience personas help you:

- Create more impactful, relevant content
- Align your team around shared goals and priorities
- Identify what formats, channels, and messaging actually move the needle

You'll also see how AI can accelerate your persona development by helping you:

- Research audience behavior and sentiment at scale
- Identify patterns and segment your audience strategically
- Draft detailed, distinct personas that guide content creation and decision-making

If you're tired of writing content that gets modest clicks and little engagement—or you're scaling a team and need to get everyone aligned fast—personas are your next power-up.

Let's uncover who your audience really is... and how to speak directly to them.

How Audience Personas Help You Create More Impactful, Relevant Content

It's easy to fall into the trap of creating content that sounds good but doesn't actually connect. Without a clear understanding of who you're speaking to, your messaging becomes too generic, your tone inconsistent, and your formats misaligned with how your audience consumes information.

Audience personas change that.

Personas are detailed, research-informed profiles of the people your content is designed to serve. They give your team a shared language and lens to create content that resonates—whether you're crafting a blog post, planning a webinar, or briefing AI on your next content campaign.

Here's how they support content creators:

1. Message-Market Fit

Personas help you zero in on what matters most to your audience. When you understand their pain points, goals, and objections, your content becomes more than informative—it becomes outrageously useful.

For example, instead of "5 Tips for Better Email Marketing," a persona-driven headline might become "5 Email Automation Tips for SaaS CMOs Looking to Reduce Churn."

2. Strategic Alignment

With personas in place, every team—marketing, sales, product—can rally around the same audience understanding. It becomes easier to prioritize content ideas, evaluate requests, and say "no" to random acts of content that don't serve your people. The result: fewer internal debates and more content that drives results.

3. Empathy-Driven Creativity

When creators truly understand their audience's motivations, tone shifts from robotic to real. Whether you're writing for a skeptical CFO or an overwhelmed HR manager, personas help you tailor your language, tone, and format to build trust more effectively. Activate empathy by adding empathy triggers to your persona doc—like what stresses them out at work or what makes them feel like a hero.

4. Smarter Repurposing

Personas make it easier to repurpose content intentionally by adapting it to each segment's needs. You can spin one core piece into several variations—each with the right angle, format, and channel based on audience preferences.

For example, a long-form research report becomes a sales-enablement one-pager for your enterprise persona, a webinar for your practitioner persona, and a podcast interview for your industry influencer persona.

5. Better AI Outputs

Let's be real: AI can only write to your audience if you define that audience for it first. Well-crafted personas are the missing context that transforms generic AI content into targeted, brand-aligned storytelling. Use persona descriptions directly in your prompts to guide tone, empathy level, and use of industry language.

Personas aren't just helpful; they're foundational. They shift content from guessing to serving. And in a noisy digital landscape, relevance isn't optional—it's your competitive edge.

Aligning Your Team—and Your Content—Around Shared Goals

One of the biggest challenges in content operations isn't creativity—it's alignment. Different people often have different goals, opinions on messaging, and preferences for what content gets prioritized. Without shared clarity, your content calendar becomes a battlefield of competing agendas or, worse, a dumping ground for disconnected content.

Audience personas bring everyone back to center.

When you use personas as a strategic anchor, your content planning shifts from "What do we want to say?" to "What does our audience need to hear?"

Here's how personas help you get your team on the same page—and focus your content on what actually moves the needle.

Personas Create a Shared Language for Prioritization

When each piece of content is tied to a specific persona, it's easier to justify—and challenge—ideas. Instead of debating whose request is more important, you can evaluate ideas based on alignment with persona needs, ROI potential, or campaign goals.

Instead of deciding whether or not to create content based on available resources, it shifts the decision point to what the content user needs. "Does this white paper address a pain point our IT Admin persona is actually struggling with—or is this just an internal priority?"

This keeps the conversation strategic, not subjective.

Personas Match Formats and Channels to Behavior

Not every persona consumes content the same way. Executives may prefer summaries and thought leadership, while practitioners look for templates or how-to guides. Segmenting your audience helps you align content formats and distribution strategies with actual audience behavior.

Ask:

- Does this persona prefer quick hits or deep dives?
- Do they spend time on LinkedIn, YouTube, email, or podcasts?
- Are they likely to attend webinars or live events?
- Will they stream audio or video content or download guides?

Personas Focus Messaging on What Matters Most

Different personas require different messages—even if you're selling the same product. A CFO wants ROI. A department lead wants efficiency. A practitioner wants ease of use. Personas help you align copy and campaigns to real motivations.

For example:

- Finance persona messaging might emphasize cost savings and long-term value.
- Operations persona content could focus on speed, scalability, or simplicity.
- Customer experience persona content highlights customer impact and real-time feedback.
- And so on...

Pair each persona with two to three message pillars they care about most. Use this in briefs and campaign planning to keep content aligned.

Personas Keep Everyone Focused on a Unified Audience Journey

Personas don't just help you define who you're talking to—they clarify when to talk to them, and how.

When your team aligns on audience segments across the funnel, you can:

- Build smarter nurture sequences
- Plan cohesive omnichannel campaigns
- Hand off leads more effectively between marketing and sales

No more one-size-fits-nobody campaigns and no more generic content!

Personas don't just support better content—they support better collaboration. When your team shares a clear view of who you're trying to reach, what they care about, and how they engage, every piece of content becomes a strategic asset, not just another asset in the marketing automation system.

How AI Can Accelerate Persona Development

Traditionally, building strong audience personas takes weeks of interviews, surveys, and behavioral analysis. It's valuable—but time-consuming. Now, with the help of AI, you can supplement that qualitative research with powerful, large-scale insights in a fraction of the time.

AI doesn't replace the human nuance in persona building. But it does amplify your ability to uncover patterns, validate assumptions, and go deeper into how your audience thinks, feels, and acts—especially across digital channels. It gives you a massive head start when building or refreshing personas. It helps you surface what your audience is already saying, so you can create content that speaks to them—not just about them.

Here's how to use AI to quickly surface behavioral and sentiment insights that strengthen your personas:

1. Analyze Social Media Conversations + Community Forums

AI tools like ChatGPT (with plugins), Claude, or custom-trained models can scan massive volumes of public conversations across platforms like Reddit, X (formerly Twitter), LinkedIn, and product review sites. This is powerful because it allows you to identify:

- Common frustrations, pain points, or "wish list" features
- The exact language your audience uses to describe their needs
- Emerging trends or unmet expectations in your space
- Sentiment patterns across different persona segments (e.g., execs vs. practitioners)

Here's a simple prompt example:

> "Summarize common pain points expressed by HR professionals in LinkedIn posts from the past six months related to employee onboarding."

Personally, I've been a long-time user of SparkToro for providing context for my personas. They make it incredibly easy to keep an eye on your social audience's activity online, explore different keywords and topics of relevance, and learn more about the audience that's interacting with that content online.

2. Review Calls, Meeting Transcripts, and Help Desk Logs at Scale

AI transcription and summarization tools (like Otter.ai, Gong, People. ai [a former client], or Fireflies) make it easy to analyze internal voice-of-customer data—from sales calls, onboarding sessions, support chats, and more.

Use AI to look for patterns in:

- Frequently asked questions
- Objections or hesitations before buying
- Emotional cues around product use or pain
- Recurring success themes or "aha" moments

Then use that language to craft your personas' goals, challenges, and emotional drivers with real-world specificity.

Here's a prompt example:

> "Summarize the top five customer challenges and emotional pain points mentioned across these transcripts. Highlight direct quotes that illustrate each theme and suggest how they could inform the goals and barriers of a buyer persona."

3. Cluster Survey Responses and Open-Ended Feedback

Do you have customer survey responses, quiz data, customer interviews, or employee feedback in spreadsheets? Feed that data to AI and ask it to cluster insights by sentiment, topic, or persona type.

AI can help you:

- Group feedback by theme or experience
- Highlight key motivators or blockers by role
- Distill long-form answers into persona-ready summaries

Here's a prompt example:

> "Analyze the following open-ended survey responses from marketing managers. What are the top recurring themes and emotional drivers?"

If you don't have current survey data, consider creating a quiz or an interactive poll to help you gain some audience insights.

"Quizzes are an interactive way to provide [customers] information while also receiving something as an organization, which is something we've been missing for a while as marketers—and why quizzes, to me, as a content marketer, are so juicy, because they give us a real connection with those folks and allow us to give as good as we get," says Maureen Jann in her #ContentChat episode[8] on interactive content.

4. Extract Intent Data from Search and Site Behavior

With access to keyword data, website analytics, or content engagement reports, AI can help you spot trends tied to specific segments.

You can uncover:

- Which personas engage with what topics
- Drop-off points in content journeys
- Gaps between high-interest topics and current content coverage

Use these insights to tailor persona pain points and preferred formats.

Here's a prompt example:

> "Analyze this website analytics and keyword report. Identify which topics different visitor segments engage with most, where they drop off in their content journey, and any gaps between high-interest topics and our existing content. Summarize the insights as pain points and preferred content formats for each persona."

5. Turn Raw Data Into Draft Personas—Fast

Once you've gathered behavioral and sentiment signals and worked through the above prompts, you can prompt AI to generate initial persona drafts for refinement.

---

8   Jann, Maureen. "#ContentChat An Introduction to the Most Underrated Content Marketing Format—The Quiz." Erika Heald Marketing Consulting, July 15, 2024. https://erikaheald.com/content-chat-why-quizzes-are-a-valuable-content-marketing-investment/. Accessed September 29, 2025.

Here's a prompt example:

> "Based on this research [paste or link data], create a persona for a mid-career IT director at a mid-sized SaaS company. Include their role, goals, content preferences, tone sensitivities, and how we can support them through content."

Drafting Detailed, Distinct Personas That Guide Content Creation and Decision-Making

Once you've gathered insights—whether through interviews, surveys, internal conversations, or AI-fueled research—the next step is to translate those patterns into usable personas.

Not just pretty slide decks. Not just aspirational marketing avatars. Practical, content-driven profiles your team can rely on every day to make better decisions, faster.

A strong persona isn't just about demographics or job titles. It's about motivation, mindset, and how your content can help them achieve something meaningful.

"A lot of us, even as seasoned marketers, think of buyer personas or target audiences often just from a demographic standpoint," says Kate DiLeo. "I challenge my clients to create a list of what I call buyer criteria, usually 20 or more, that include psychographics. Let's say you sell software to B2B. Okay. Who cuts the check? What are the biggest pains? What's the level of bureaucracy? Are the teams aligned or not aligned? How do they communicate? Are they autonomous, collaborative, and coachable?"

With Kate's advice in mind, here's how to bring your personas to life in a way that drives action—not just awareness:

Focus on Usefulness, Not Fiction

The goal isn't to write a character study of Sales Pro Sam. Your persona should feel like someone your team could talk to in a meeting tomorrow.

Include only the details that help you create tailored content, make strategic content decisions, and serve their needs more effectively.

Skip details like "enjoys cold brew" or "has two children" unless it meaningfully affects how or where you connect with them.

Use a Practical, Repeatable Format

Give each persona its own one-pager with consistent fields that allow for comparison and easy reference. Here's a format to start with:

- **Persona name**. Use a real first name, not a cute label like "Tech Tina."
- **Role/Title(s)**. List the job titles this persona typically holds.
- **Industry/Company Size**. Note the sectors and organization sizes they're most often part of.
- **Decision-making role**. Clarify whether they're a decision-maker, influencer, blocker, or other stakeholder.
- **Key goals**. Capture what this persona is working to achieve in their role.
- **Core challenges or frustrations**. Outline the main obstacles or pain points they face.
- **What success looks like for them**. Define the outcomes or achievements they would consider a win.
- **What they value in content**. Specify what makes content useful for them (e.g., data, brevity, case studies).
- **Preferred content formats and communication channels**. Identify how and where they like to consume content (e.g., email newsletters, LinkedIn, podcasts).
- **Messaging that resonates**. Highlight key phrases, proof points, or tones that connect with this persona.
- **Representative quote**. Add a real or realistic statement to humanize the persona.

Make Each Persona Clearly Different

You don't necessarily need a dozen personas. You need a manageable set of distinct ones that clearly represent different segments of your actual audience. Ask yourself, would this persona require a different message, call to action, or content format? Would they need different nurturing paths or conversion points? If not, they might be variations— not distinct personas.

Start with two to four core personas and expand as needed. Use audience behavior and your data analysis to drive additions—not guesses.

Tie Personas to Content Strategy + Editorial Planning

Personas are only valuable if they actively influence your content decisions. To get the most impact, connect them directly to your planning and production workflows rather than letting them sit unused in a slide deck.

One way to apply personas in practice is by including a "Target Persona" field in every content brief, ensuring each piece is created with a clear audience in mind. You can also build campaign maps that outline what each persona needs at different stages of the funnel so your team delivers the right content at the right time.

Personas should guide how you prioritize channels, formats, and tone across deliverables, keeping your content consistent and audience centered. Finally, tailoring AI prompts with persona traits can help generate more relevant and authentic output, bridging strategy with execution.

Make Personas Easy to Find—and Use

Effective personas aren't about having more—they're about having clarity. A few well-researched, distinct personas can unlock more focused, efficient, and empathetic content creation across your entire business.

When your team uses personas every day—whether they're writing a subject line, designing a landing page, or prompting an AI tool—that's when your content starts to sound less like noise... and more like exactly what your audience needed to hear.

AI Prompts for Creating Useful Personas

Use these prompts as is or modify them based on your generative AI tools, access to internal data, and business goals. These templates come directly from asking ChatGPT how to write prompts for the use cases mentioned in this chapter.

Researching Behavior + Sentiment from Public

Prompt 1: Compare Persona Segments by Platform

"Analyze public conversations about [TOPIC] across Reddit, LinkedIn, and product reviews. Compare how executives vs. practitioners discuss the issue and highlight differences in tone, concerns, and goals."

Prompt 2: Spot Emerging Needs Before They Trend

> "Review the last three months of discussions on [FORUM/ PLATFORM] related to [TOPIC]. Identify early signals of unmet needs, frustrations, or new priorities that could shape persona updates."

Analyzing Internal Data at Scale (Surveys, Support Logs, Interviews)

Prompt 3: Translate Feedback Into Customer Journeys

> "From this collection of onboarding call transcripts, cluster feedback into journey stages (awareness, consideration, purchase, onboarding, renewal). For each stage, note common pain points and success signals."

Prompt 4: Detect Emotional Language in Support Logs

> "Analyze these anonymized support tickets. Highlight recurring emotional cues (e.g., urgency, frustration, relief) and link them to persona challenges and priorities."

Drafting Persona Profiles from Synthesized Data

Prompt 5: Contrast Two Persona Archetypes

> "Based on this dataset, create a comparison table for two primary personas. Show differences in role, goals, barriers, and preferred content. Include example language each might use."

Prompt 6: Identify Content Tone Sensitivities

> "From this transcript data, identify words or phrasing that land positively vs. negatively with [PERSONA ROLE]. Suggest tone guidelines for writing content that resonates."

Using Personas to Inform Content Briefs and AI Content

Prompt 7: Prioritize Topics by Persona Value

> "Given this list of blog topics and performance data, rank the top five for [PERSONA ROLE] based on alignment with their goals, pain points, and engagement patterns."

Prompt 8: Create Persona–Specific FAQs

"Generate a list of 10 FAQs for [PERSONA NAME] about [TOPIC]. Use language they would actually use and note which funnel stage each question maps to."

Persona-Powered AI Content Creation Prompts

Prompt 9: Adapt Content for Different Buying Stages

"Take this draft article [PASTE TEXT] and rewrite it three ways:

1. For Awareness stage [PERSONA NAME]

2. For Consideration stage [PERSONA NAME]

3. For Decision stage [PERSONA NAME]

Highlight shifts in tone, CTA, and proof points."

Prompt 10: Extract Persona–Aligned Headlines

"Generate five alternative headlines for this draft article [PASTE TEXT], each tailored to a different persona segment: [PERSONA A], [PERSONA B], [PERSONA C]."

Persona Maintenance + Research

Prompt 11: Benchmark Against Competitors' Personas

"Analyze the messaging from these competitor websites [URLs]. Based on tone, content themes, and CTAs, infer what personas they may be targeting. Suggest updates to differentiate our personas and content."

An Example Content Persona

This persona is a version of one that I've generated with help from AI and have used to guide my own work.

Persona Snapshot

Name: Maria Lopez

Role/Title: Director of Content Marketing

Industry/Company Size: B2B SaaS, ~300 employees

Decision-Making Role: Decision-maker for content strategy; influencer in marketing tech stack purchases

Representative Quote: "If my team can't show measurable impact on pipeline and brand trust, we risk being seen as a cost center instead of a growth driver."

Priority Initiatives (budget & decisions)

1. Investing in content management systems and AI tools that improve speed and scalability.

2. Building and maintaining a content governance framework (style guide, voice chart, templates).

3. Developing thought leadership programs with executives and SMEs.

4. Expanding multi-channel distribution (LinkedIn, industry newsletters, partner webinars).

5. Funding research and data-driven assets (original reports, case studies).

6. Supporting SEO and content repurposing to maximize ROI.

7. Training internal teams and onboarding freelancers/contractors.

8. Sponsoring select industry events and content syndication partnerships.

9. Testing new content formats (interactive tools, video, podcasts).

10. Strengthening collaboration with sales enablement and customer success teams.

Success Factors

- Consistently publishing high-quality, on-brand content on schedule.
- Demonstrated contribution to lead generation and pipeline velocity.
- Positive feedback from sales teams on the usefulness of content assets.
- Improved search visibility and organic traffic growth.
- Reduced content production bottlenecks via governance systems.
- Growth in executive visibility through thought leadership programs.

- Recognition as a trusted advisor internally and at industry events.
- Team retention and satisfaction (avoiding burnout).

Perceived Barriers

- Limited headcount: too many demands on a small team.
- Executive stakeholders who underestimate the time required for content excellence.
- Constant ad hoc requests ("random acts of content") that derail priorities.
- Difficulty accessing SMEs for timely input.
- Overlap or conflict with product marketing and PR priorities.
- Lack of consistent metrics or dashboards for proving ROI.
- Budget scrutiny when immediate revenue attribution isn't clear.

Day in the Life

- **6:30 am**: Checks Slack and email on her phone before heading to the gym; scans industry newsletters (Content Marketing Institute [CMI], MarketingProfs, Digiday).
- **8:30 am**: Logs in at home office; reviews editorial calendar in Notion and ensures today's deliverables are on track.
- **9:00 am**: 1:1 with a content strategist; discusses AI tool outputs and editorial priorities.
- **10:30 am**: Joins cross-functional meeting with sales enablement; aligns on upcoming product launch collateral.
- **12:00 pm**: Quick lunch while catching up on LinkedIn thought leadership posts from peers (Robert Rose, Ann Handley).
- **1:00 pm**: Reviews and edits drafts in Google Docs; provides feedback on tone and alignment with persona needs.
- **3:00 pm**: Prepares marketing slides for the quarterly leadership meeting, focusing on KPIs and campaign ROI.
- **4:30 pm**: Syncs with her small team; addresses blockers (e.g., waiting on SME input).
- **6:00 pm**: Wraps up by updating analytics dashboards; drafts a brief for an external writer.
- **Evening**: Skims non-marketing-related Substack newsletters for inspiration.

Decision Influencers

- CMO (budget sign-off)
- Sales leaders (content adoption)
- Peers in marketing operations

Go-To Resources

- CMI
- MarketingProfs
- SparkToro
- LinkedIn posts and livestreams (#ContentChat)
- Podcasts like This Old Marketing

CHAPTER 4

Taming the Chaos with Taxonomy

What if I told you the fastest way to scale your content program wasn't more writers, a better CMS, or even AI—but a spreadsheet?

But not just any old spreadsheet, of course. This one is your taxonomy—a foundational but often overlooked content tool.

In nearly every workshop I've led on content strategy and AI-assisted marketing, one truth surfaces again and again: AI can't help you scale what it doesn't understand. And humans can't maintain quality at scale without a shared understanding of how content is organized and classified.

But too many teams skip this step. Why? Because taxonomy sounds technical. Bureaucratic. Optional. Boring.

It conjures images of librarians and Latin naming conventions, not the fast-moving world of B2B marketing. And yet, it's exactly the kind of structure your team—and your AI tools—desperately need.

Taxonomy is the science of classification. It's not sexy, but it's what gives AI and your content team the context they need to do their jobs without guesswork.

Without a defined taxonomy, you're asking AI (and your team) to work in the dark. You'll see it reflected in duplicate tags, chaotic CMS fields, and reports that show metrics for five different versions of the same topic—all spelled slightly differently. You'll feel it in the missed opportunities for personalization, the time spent searching for the right asset, and the inability to spot trends across channels.

When you take time to define and document your taxonomy, you unlock a host of strategic advantages:

- Consistent naming and tagging conventions that clean up your content universe
- Smarter AI performance thanks to clearer context
- Scalable personalization and segmentation without reinventing the wheel
- Actionable analytics that actually reflect your audience's behavior and content performance

Taxonomy provides a systematic way to understand and navigate complex information. It enables retrieval, analysis, and content personalization that's only possible with structure.

In this chapter, we'll break down what a useful taxonomy looks like in practice—not theory. You'll learn:

- Why content—not IT—needs to define the taxonomy
- How to structure a scalable, hierarchical taxonomy that makes sense for your organization
- Where and how to involve AI in the taxonomy process
- Tips to avoid the most common (and costly) tagging mistakes
- A framework for incorporating taxonomy maintenance into your regular content hygiene

You'll also see real-world examples of how something as simple as clarifying your blog tags or email categories can dramatically improve both operational efficiency and campaign performance.

Because the truth is: taxonomy is what makes scale possible. And if you want to do more with less—and to do it well—it's time to stop skipping this step.

Let's dig in.

Why Content—Not IT—Needs to Define Your Brand's Taxonomy

Taxonomy has historically been seen as the domain of IT or data teams within larger companies. And while those stakeholders absolutely need a seat at the table, they should not be leading the conversation. When taxonomy lives solely within IT, it's often designed to reflect internal systems, not how audiences consume, find, or interact with content.

That's a problem—especially for content marketers trying to scale without reinventing the wheel every time.

If the content team isn't defining taxonomy, it's relying on someone else's idea of what the content is about. And that's a fast track to disorganization, poor discoverability, and missed opportunities for reuse and personalization.

While the ownership challenges apply mostly to enterprise content programs, taxonomy's value is significant for content entrepreneurs and small businesses, too.

Taxonomy isn't about tagging for tagging's sake. It's about creating shared context. It's how we align creators, platforms, and tools—especially AI—on what content exists, what it's for, and how it fits into the larger content strategy.

When taxonomy is owned by content (with input from SEO, analytics, and IT), it reflects:

- Audience-first language—not jargon or internal team names
- Actual usage patterns across channels and content types
- Content goals and lifecycle stages, from top-of-funnel awareness to post-sale education
- Strategic alignment with campaign messaging, personas, and business objectives

That's why I always advocate for taxonomy to be led by the content function. We're closest to the intent behind the content. We understand how our audience thinks and searches. We're the ones constantly auditing what exists and identifying the gaps. This work is inherently strategic—it belongs with the team responsible for content performance.

Of course, this doesn't mean working in a silo. Strong taxonomy is always a cross-functional effort, with content in the lead and input from:

- SEO and analytics, for search behavior and reporting structure
- IT or operations, for implementation across tools and platforms
- Business stakeholders, to ensure alignment with product, brand, and customer-facing teams

But let's be clear: taxonomy is not a backend IT project. It's a core content operations function—and one of the most impactful things a content team can do to enable scale, clarity, and AI-readiness.

How to Structure a Scalable, Hierarchical Taxonomy That Makes Sense for Your Organization

A good taxonomy doesn't just help you organize content—it helps your content work harder. It ensures that assets are easy to find, simple to repurpose, and accurately matched to the right audience, channel, or objective. But to deliver on that promise, your taxonomy must be more than a list of tags. It needs to be scalable, hierarchical, and aligned with how your organization communicates—not just how your content management system wants to store information.

The good news? You don't need to start from scratch. You do, however, need to be intentional.

Start with What You Already Know

Before creating anything new, take a pause and audit:

- Your current categories and tags (and how consistently they're used)
- Your main content channels and how content is grouped within them
- Your personas or segments, and any naming conventions that reflect audience needs
- Your content calendar themes, campaign structures, or product/service groupings

Start by exporting what's in your CMS or content repository. You'll likely find duplicates (like "e-book" vs. "ebook" vs. "E-book"), inconsistencies, and gaps. That's your signal to bring order to the chaos.

Build a Hierarchical Structure

Taxonomy is inherently hierarchical—it creates a nesting structure, with broad categories at the top and more specific terms below. Think of it like your website's information architecture or a product catalog. You're guiding both humans and machines through layers of specificity.

A typical marketing taxonomy might look something like this:

- Channel
  - Social Media
    - LinkedIn
    - Instagram
  - Email
    - Newsletter
    - Triggered Campaigns
- Content Type
  - Blog Post
  - E-book
  - Webinar
- Persona
  - IT Decision-Maker
  - Mid-Market CMO
- Journey Stage
  - Awareness
  - Consideration
  - Decision

You don't need to build this all at once, and you don't need to go 10 levels deep. Start with a few high-value dimensions that reflect how your team creates and categorizes content.

Tie It to Strategic Goals

A scalable taxonomy should reflect the strategic questions your team asks regularly:

- What kind of content performs best for X audience?
- Which formats convert at Y stage of the funnel?
- Where do we have gaps in messaging or coverage?

If your taxonomy can't help you answer those questions, it's time to rework it.

Make Your Taxonomy Scalable for AI and Humans

As you structure your taxonomy, remember: you're designing for both human editors and machine readers. AI tools don't intuitively know that "case study" and "customer story" are the same thing unless you tell them. The clearer and more structured your taxonomy, the better your AI tools (and analytics platforms) will perform.

When in doubt, lean into naming clarity over cleverness, and document every decision so all brand content creators are using the same terminology.

Where and How to Involve AI in the Taxonomy Process

AI can be a valuable partner in taxonomy work—but only when it's used intentionally. It won't define a content taxonomy for you from scratch. And it certainly won't intuit what matters most to your brand or audience. But with the right prompts and context, it can accelerate content cleanup and discovery in ways that would take humans hours (or weeks) to uncover manually.

Think of AI as a collaborator—not a strategist. It doesn't replace your judgment; it reflects it back to you, faster.

Where AI Adds Value for Your Taxonomy

Here are some smart, strategic points to bring AI into your taxonomy development:

Auditing Existing Content Tags

If your blog, CMS, or DAM is cluttered with overlapping or inconsistent tags, AI can help:

- Surface variations (e.g., "e-book" vs. "ebook" vs. "eBook")
- Group similar tags together
- Flag terms that are unclear or redundant

You can do this with a simple export and a prompt like:

> "Here's a list of 300 tags used across our website. Group them by similarity and suggest standardized tag names."

Inferring Structure from Published Content

If you're not sure how your content breaks down by format, funnel stage, or persona, you can ask AI to help identify patterns in tone, topic, or layout—especially if you paste in representative examples or link to indexed pages.

Here's an example prompt:

> "Analyze these 20 blog posts. What primary topics, audience types, or stages of the customer journey do they seem to target?"

This approach helps you reverse-engineer structure from what's already working.

Suggesting Hierarchical Labels

Once you've drafted broad categories, AI can help you build logical subcategories or suggest nested relationships. You'll want to review these for brand fit, but it's a quick way to generate ideas you might not have considered.

Here's an example prompt:

> "Here are 10 high-level content types we use. Suggest two to three subcategories for each, based on common B2B content taxonomies."

Helping with CMS Cleanup

When paired with a spreadsheet export from your CMS, AI can assist with mass reclassification, flagging duplicates, or creating rules for redirects or merges. You'll still need a human to approve final decisions, but AI makes the prep work lighter.

What AI Can't Do[9] (And Why It Matters)

AI can't:

- Prioritize based on what's most important to your business
- Understand your audience's nuanced expectations
- Align taxonomy to campaign goals or sales stages
- Recognize brand-specific language unless explicitly trained

---

9 This isn't my opinion, it's an edited version of what ChatGPT told me it's not good at.

That's where your content strategy comes in. You bring the life experience and judgment. AI brings speed.

The best outcomes happen when you pair structured guidance (like templates, prompts, and strategic parameters) with human oversight. So, before you bring AI into the taxonomy process, make sure you've already clarified your:

- Strategic goals for taxonomy
- Key content types and channels
- Audience personas and funnel stages

Otherwise, you're just asking AI to organize the chaos—and you'll still end up with a mess.

Tips to Avoid the Most Common (and Costly) Tagging Mistakes

A well-structured taxonomy is only as useful as its implementation. And implementation is where many teams fall into trouble—not because of bad intentions, but because of a lack of clarity, consistency, or accountability. Even a thoughtfully built taxonomy can spiral into chaos if your tagging practices aren't governed.

Let's talk about the common mistakes that derail taxonomy—and how to avoid them from the start.

1. Letting Everyone Tag Content Freely

This is the fastest route to taxonomy bloat. When everyone can create their own tags ad hoc—without guardrails or review—you end up with dozens of near-duplicates and categories no one understands.

Think:

- "customer success story" vs. "case study"
- "blog" vs. "blog post" vs. "article"
- "mid-market buyer" vs. "midmarket buyer" vs. "SMB"

How to fix it: Restrict tag creation permissions in your CMS or DAM. Only allow additions through a documented request process. Assign a taxonomy owner or editor to review all new suggestions against your existing structure.

2. Creating Tags That Are Too Broad (or Too Specific)

Tags like "marketing" or "content" may sound useful, but they don't actually help anyone find relevant content. On the flip side,

hyper-specific tags like "Q2 2025 Instagram Paid Campaign – Retail" won't scale and may only apply to one asset.

How to fix it: Create a sweet spot for tags—specific enough to be meaningful, broad enough to apply to more than one piece. Consider combining them with categories, subcategories, and metadata fields for granularity instead of trying to cram everything into one tag.

3. Using Inconsistent Naming Conventions

Capitalization, punctuation, and word order matter. Your CMS might treat "E-Book," "e-book," and "ebook" as three different things

—and guess what? So will your analytics tools and AI.

How to fix it: Document and enforce a clear naming standard (e.g., lowercase, hyphenated, singular nouns). Use data validation or dropdown menus whenever possible to enforce consistent inputs.

4. Forgetting About the End Users (Internal and External)

It's easy to build a taxonomy that makes sense to the content team—but completely fails for sales enablement, SEO, or customer self-service. If your internal users can't find what they need, they'll duplicate work. If customers can't navigate your resource library, they'll bounce.

How to fix it: Test your taxonomy with real users. Ask a marketer, seller, or customer success rep to find content using your categories. Watch where they get stuck and revise your tagging accordingly.

5. Ignoring Taxonomy Maintenance

Even the best taxonomy needs ongoing care. Content gets outdated. Business priorities shift. Tags become irrelevant or unused.

How to fix it: Build regular taxonomy audits into your content hygiene workflow (we'll cover that in the next section). Tagging isn't a one-and-done project—it's part of running a scalable content operation.

Avoiding these mistakes isn't about being perfect—it's about being proactive. Taxonomy isn't just metadata. It's infrastructure. And when that infrastructure is shaky, everything built on top of it suffers.

Incorporating Taxonomy Maintenance Into Your Regular Content Hygiene

Taxonomy isn't a set-it-and-forget-it project. It's a living, operational asset that evolves with your business. If left unattended, even the most well-structured taxonomy will start to decay—leading to inconsistent tagging, unusable filters, and AI that suddenly "forgets" how to contextualize your content.

The key to keeping it useful? Treat taxonomy like any other critical content system: it needs governance, maintenance, and checkpoints.

Here's a framework you can adopt (or adapt) based on your team size and capacity:

Monthly: Spot-Check for Consistency

What to review:

- New content added since the last review
- Tags applied by external contributors or freelancers
- Deviations from naming conventions

How to do it:

Review new content in batches, or set up a basic workflow where your taxonomy owner checks new entries weekly or monthly. If you have a content ops tool, use automations to flag any rogue tags.

Quarterly: Audit for Relevance and Redundancy

What to review:

- Tags or categories that are no longer in use
- Duplicate or near-duplicate terms
- Tags with only one piece of content attached
- "Other" or "miscellaneous" categories (often a red flag!)

How to do it:

Export your content database or CMS tagging fields. Sort by usage volume and date last applied. Sunset or merge anything that no longer serves a strategic purpose.

This keeps your taxonomy lean—and your analytics clean.

Annually: Reassess for Strategic Alignment

What to review:

- Does your taxonomy still reflect your core personas and funnel stages?
- Are there new content types, business units, or messaging
- themes that should be added?
- Have any tags become irrelevant due to market or organizational shifts?

How to do it:

Bring your core content stakeholders together for a taxonomy working session. Review performance data and user feedback. Update your taxonomy documentation to reflect any changes—and re-share it with all content creators and collaborators.

Ongoing: Embed the Taxonomy Into Your Workflows

As with most content tools we've talked about in this book, taxonomy benefits from being integrated into your workflows.

How to do it:

- **Template it**. Your content briefs, intake forms, and publishing checklists should all link to or include the approved taxonomy.
- **Train for it**. Include taxonomy standards in your onboarding for content creators, freelancers, and agency partners.
- **Tool it**. Use drop-downs, required fields, and automation wherever possible to reduce human error and tagging inconsistencies.

The goal here isn't perfection—it's sustainable clarity. With a lightweight, repeatable hygiene routine, you'll avoid the common fate of most content repositories: bloated, outdated, and impossible to search.

A well-maintained taxonomy keeps your content ecosystem usable—for humans, AI, and your bottom line.

Implementing Streamlined Processes and Templates

Part I of this book was about defining who you are, what you stand for, and how that shows up in your content. Part II is about building the systems to deliver on that promise—consistently, efficiently, and at scale.

Developing scalable content programs or content entrepreneurship isn't just about creating great content. It's about creating it strategically

—in ways that protect your time, amplify your impact, and grow your business without burning you out.

That means moving beyond just winging it and embracing the unsexy (but wildly effective) tools of process documentation, templates, and workflows.

In this section, you'll lay the foundation for a sustainable content operation that evolves with your business. You'll learn how to:

- Document what works so you don't reinvent the wheel every time
- Delegate effectively—whether to AI, freelancers, or future teammates
- Build repeatable systems that reduce mental load and maximize creativity

These chapters will give you real-world examples, customizable templates, and repeatable processes I've developed and tested in both in-house and consulting roles. If you're ready to stop overthinking and start streamlining, this is your playbook.

This section is for anyone who's tired of improvising and ready to create content systems that actually support the growth you've been working toward.

Let's get your content operation running like the strategic engine it's meant to be.

CHAPTER 5

Refining Content Ideation and Planning

Let's be real: Content planning can be overwhelming—especially when you're wearing more than one hat in your business.

Brainstorms turn into rabbit holes. Idea lists get bloated or abandoned.

Planning sessions start with good intentions... and end with decision fatigue.

But it doesn't have to be that way.

When you have a repeatable process for ideation and planning, you shift from reactive scrambling to intentional, strategic creation. You stop wasting energy rethinking how you'll plan—instead, you focus on what matters most: making content that serves your audience and supports your business goals.

In this chapter, we'll take the mystery and mess out of content planning with:

- Templates that give shape to your brainstorming sessions and guide your content creation
- Processes that are easy to replicate to reduce friction and overthinking
- Tools that support faster research and smarter topic selection
- Systems that help you prioritize ideas that actually align with your goals

You'll learn how to build a sustainable content planning workflow that works whether you're a solo creator or starting to scale your team.

Because the real magic of content entrepreneurship isn't endless ideas—it's having a plan that turns the right ideas into content that works.

The Value of Structured Content Templates

Structured content templates do more than just speed up content creation—they bring order, repeatability, and quality control into what can otherwise be a chaotic process. Without templates, content creation becomes a high-effort, high-friction experience every single time.

Here's why I always use structured templates:

They reduce decision fatigue

No one should start content creation with a blank screen. Templates help focus the creator by explaining:

- What kind of content are we creating?
- What does success look like?
- What sections must be included—and in what order?

This mental scaffolding reduces the time spent figuring out how to begin and keeps creators focused on the job the content needs to do.

They enable better collaboration

When everyone uses the same framework, expectations are aligned. Freelancers, SMEs, and internal stakeholders all have a shared map of what's needed—reducing rework, review cycles, and missed deadlines.

Templates let you scale and collaborate more effectively so you're not the only one who can move things forward.

They protect brand consistency

Templates help reinforce voice, tone, formatting, and structural conventions—especially when paired with guidance on how to use them correctly.

They make governance operational

Your content governance guidelines (brand voice, SEO, legal disclaimers) only work when they're implemented consistently. Templates serve as the delivery mechanism for governance, improving efficiency and productivity.

Template Best Practices

I have an unpopular opinion that I hope to sway you on: Templates aren't creativity killers—they're clarity enablers. When built thoughtfully, templates don't box content creators in; they free them up to focus on storytelling, subject matter, and audience needs, rather than having to remember formatting rules or stakeholder requirements from scratch.

Still, not all templates are created equal. A bloated, overly complex template can cause as much unproductive friction as no template at all. So, how do you create templates that are actually used, loved, and effective?

Here's what to keep in mind:

Standardize the strategic alignment

Every template should open with the same few prompts that connect the piece back to strategy:

- Objective: What is this content supposed to accomplish?
- Audience: Who is it for, and what do they need?
- Journey Stage: Where are they in their decision-making?
- CTA: What do we want them to do next?

Use the same structure and format across your entire template library so that contributors know what to expect. Repetition helps teams internalize your process and dramatically reduces friction.

This strategic alignment ensures that even one-off or "quick" content still ladders up to business goals.

Include links to examples and resources

Don't assume users will know what "good" looks like. Link out to:

- High-performing examples
- Your content style guide

- Your brand voice chart
- SEO guidelines or metadata fields

The context these links provide speeds up content creation and reduces rounds of review.

Tailor templates by channel and content type

What's required for an email is very different from what's needed for an e-book or LinkedIn post. You don't need dozens of templates—but you do need purposeful ones for your most frequent content types and channels.

You shouldn't rely on people remembering all the must-includes for your different formats. Your templates should do that for them.

Make every template fill-in-the-blank friendly

Use checkboxes, dropdowns, and field prompts wherever possible. Make it as easy as possible for a stakeholder or subject matter expert to provide the input your content team needs.

A great content template isn't just a place to jot down ideas. It's a strategic blueprint that guides every step of your content creation process—from planning through publication. And once it's built, it pays dividends every time you create a new draft.

How Templates Help You Scale

When you're a content team of one (or even a team of 10), scaling content without templates is like baking without recipes: messy, inconsistent, and time-consuming. Templates are the key to transforming content creation from an ad-hoc scramble to a replicable, efficient engine. Here's how they unlock scale:

They cut production time

Templates remove guesswork and reduce the need for repeated explanations. New content types can be produced faster, with less oversight—even by less-experienced contributors.

They streamline repurposing

When content is structured from the start, it's easier to reuse and adapt. You can take one high-performing asset and quickly spin it into:

- A webinar landing page
- A social media series
- An email nurture sequence
- A short-form video script

They improve quality control

Fewer missing metadata fields. Fewer overlooked CTAs. Fewer urgent "Oh no, we forgot to link to the case study!!!" texts while you are on vacation. Templates catch these gaps early, reducing the burden on editors and reviewers.

They make it easier to measure what matters

When your content follows a consistent structure, it's easier to track performance across formats, channels, and campaigns. You can compare like with like—and act on the data.

Without templates, your team has to do everything from scratch every time. With them, you have reusable blueprints that make everything easier—from the first draft to the final review.

10 Essential Templates to Power Your Full Content Workflow

These are the 10 most important templates I use to transform content creation from guesswork into a repeatable, scalable system—from ideation to publication.

Content Planning & Brief Template

This is the backbone of your content creation process. It gives every piece a clear reason to exist, defines who it's for, and sets expectations for execution and measurement. A well-built content brief doesn't just prevent scope creep—it keeps creators focused, stakeholders aligned, and content tied to your marketing goals from day one.

The brief should always start with strategic context. Before anyone starts writing—or even brainstorming—your team should understand why the content matters and who it's meant to serve.

Strategic Foundation (Top Section)

This part grounds the piece in your content strategy. Every brief should clearly identify:

- **Target Persona**

  - Who is this content for? Reference your documented persona or audience segment.

- **Business Goal**

  - What is this content meant to achieve? Brand awareness? Lead gen? Product education?

- **Customer Journey Stage**

  - Where in the buyer's journey does this content fit? Awareness, consideration, decision, or retention?

- **Content Type**

  - Blog post, case study, email series, white paper, video, etc.

- **Proposed Title or Angle**

  - What's the working headline or narrative hook?

- **Primary CTA**

  - What should the reader do next? Is there related content they should be offered?

Execution Details (Middle Section)

This is where the brief gets tactical—but still structured for clarity.

- **Key Talking Points or Outline**

  - A high-level outline or list of core messages to include.

- **Subject Matter Expert (SME)**

  - Who's providing insight or review?

- **Research or Source Links**

  - Include background docs, interviews, or supporting data.

- **Visuals or Multimedia Needs**

  - Note required images, video clips, charts, or pull quotes.

- **SEO/GAO**

  - Include target keywords, question to answer, and a TL;DR summary if applicable.

Logistics & Workflow (Bottom Section)

This final section ensures accountability and smooth collaboration.

- **Due Date & Publish Date**

  - Be realistic—and aligned with your content calendar.

- **Distribution Plan**

  - Where will this live? What channels will support it?

- **Review Steps**

  - Who needs to approve this and in what order?

- **Final Checklist**

  - Add a checklist for things like voice alignment, metadata, compliance, and accessibility.

Structure Tip

Begin your template with the strategic why before diving into execution. Too many content briefs focus solely on word count or keywords, missing the opportunity to align everyone around the bigger picture. Your goal is to create content that's not just published—but purposeful.

Brainstorming & Topic Capture Worksheet

This worksheet bridges the gap between raw ideas and real strategy. It's where sparks of inspiration get captured, shaped, and assessed—without immediately demanding a full brief or editorial slot. It gives your team a lightweight, repeatable way to record content ideas and start evaluating their potential.

Not every idea captured here will become a published piece—and that's okay. The goal is to build a backlog of thoughtful, audience-first concepts that can be prioritized based on relevance, alignment, and business value.

What to Capture

This worksheet works best as a spreadsheet or form—something simple to keep up with in the moment. Here's what to include:

- **Working Title or Concept**

  - A headline, a phrase, or just a sticky note–style label for the idea.

- **Audience / Persona**

  - Who is this for? Align with your existing personas or target segments.

- **Problem or Opportunity**

  - What question is this answering? What pain point or curiosity does it tap into?

- **Proposed Format(s)**

  - Blog? Webinar? Video series? Short-form social post? Include more than one if the idea is flexible.

- **Tilt, Hook, and Supporting Data**

  - What's the unique take? Is there a stat, case study, or trend that makes this idea stand out?

- **Content Pillar or Theme**

  - Tie the idea to one of your core strategic buckets— like thought leadership, product education, or customer success.

- **Business Goal Alignment**

  - Is this idea designed to drive traffic? Nurture leads? Support retention? Even a quick tag here will help you prioritize later.

Structure Tip

Keep the worksheet open-ended enough to encourage creativity but structured enough to ensure ideas can be acted on. If your team can't easily tell how an idea maps to a goal or content pillar, it's probably not ready to move forward yet.

This worksheet becomes even more powerful when it's revisited regularly—during monthly content planning meetings or quarterly strategy refreshes. It's not just a parking lot for ideas—it's an on-deck circle for your editorial calendar.

Content Pillar + Subtopic Mapping Sheet

This mapping sheet helps you stay strategically aligned with your business goals while leaving room for creativity. It's your anchor for organized ideation—ensuring every piece of content ladders up to something meaningful, while also giving you visibility into what's been covered, what's in progress, and where your gaps are.

When you skip this step, you end up with scattered blog topics, one-off campaigns, and assets that don't connect or compound. When you make it part of your content workflow, you build consistency, reduce redundancy, and enable scalable repurposing.

What to Include (Core Fields)

This mapping sheet should strike a balance between structure and flexibility. Whether you're using a spreadsheet, a Trello board, or a Notion database, make sure it includes:

- **Content Pillar**

    ◦ These are your strategic themes—like Industry Trends, Product Education, Customer Success, or Leadership POVs. Typically, three to five main buckets are ideal. Make your defined pillars clear by using checkboxes or pulldowns.

- **Subtopics**

    ◦ Specific angles or ideas that fall under the broader pillar. For example, under "Customer Success," you might have "Onboarding Tips," "Retention Strategies," or "Case Studies."

- **Format Options**

  ○ What formats are best suited to this subtopic? You might note blog post, webinar, infographic, video short, etc. Standardize this by including a comprehensive list of all active publishing channels.

- **Contributor Notes**

  ○ Who on your team is the best resource for this? Are there relevant SMEs, existing assets to repurpose, or internal perspectives to highlight?

- **Status or Stage**

  ○ Optional but helpful—track whether the idea is proposed, in production, or published.

Use It to Guide, Not Limit

This sheet isn't here to dictate exactly what to publish—it's here to guide brainstorming and reduce off-strategy content creation. It helps your team see the big picture while still drilling down into practical execution.

It also becomes a powerful tool during:

- Editorial planning sessions
- Campaign alignment meetings
- Stakeholder intake reviews
- Quarterly content performance audits

Structure Tip

Use a visual format if possible. A Trello board or Notion database works beautifully here—allowing your team to filter by pillar, format, contributor, or status. The goal is to make this a living tool, not a static doc.

The mapping sheet gives everyone a shared mental model for what you talk about—and how. It's one of the easiest ways to align cross-functional teams and avoid the random acts of content trap.

Idea-to-Execution Project Tracker

This tracker is your behind-the-scenes project manager. It keeps every piece of content in motion—strategically aligned, on time, and with a clear owner. While your editorial calendar offers a top-down view of publishing cadence, this is the place where ideas are turned into executed, approved, published content.

The tracker ensures that once a concept is agreed upon, it doesn't disappear into the void. Every piece has a home, a status, and a next step. It also helps you opt out of status update meetings that should have been an email linking to an updated content tracker.

What to Track

Set this up as a spreadsheet, Airtable base, Notion database, or whatever tool your team already uses and trusts. The structure should balance strategic context with logistical clarity:

- **Working Title**

    ◦ A placeholder or draft headline to identify the piece throughout development.

- **Persona / Target Audience**

    ◦ Who are we creating this for? Reference your content personas or journey stages.

- **Pillar or Theme Alignment**

    ◦ Tie each piece to one of your core content pillars or campaign goals.

- **Content Type / Format**

    ◦ Blog post, case study, landing page, video, webinar, etc.

- **Assigned Owner**

    ◦ The person responsible for moving the piece forward. Could be a writer, strategist, or SME.

- **Status**

    ◦ Clear stages like: Idea | Drafting | In Review | Final | Scheduled | Published. Customize based on your workflow.

- **Due Date**

  - The internal deadline for the final draft or stakeholder review.

- **Publish Date (Optional)**

  - If scheduled, note when the piece is expected to go live.

- **Review or Approval Notes**

  - Capture any required checkpoints (e.g., legal, exec, brand) or final checklist confirmations.

Structure Tip

Treat this as a living document. This isn't something you update once a month—it should reflect real-time movement through your workflow.

That's where tools like Airtable or Notion shine, especially when paired with automations, like status-triggered reminders or Slack pings.

Even better? Integrate it with your content brief template so ideas can flow directly into this tracker once approved—eliminating duplication and keeping strategy intact.

Content Draft Template

This template helps content creators move from strategy to storytelling with confidence. It merges the clarity of a brief with the structure of a first draft—so writers don't have to toggle between two documents or guess at what matters. Instead, they start aligned on voice, audience, and purpose and flow straight into writing.

Unlike one-size-fits-all templates, this one is built for the specific content type. Your blog post template will look different from the one you create for your YouTube script or LinkedIn livestream events. Each template must reflect the unique structure, tone, and creative needs of the channel it supports.

What to Include

While each content format has unique needs, your core content draft template should include:

Strategic Alignment (Top Section)

This sets the direction before a single sentence is written.

- **Working Title**
  - Flexible but focused; anchors the theme.
- **Synopsis or Purpose Statement**
  - What this content is about and what it's meant to achieve.
- **Target Audience**
  - Which persona this is for and where they are in the journey.
- **Primary CTA**
  - What action we want the reader/viewer to take.
- **Pillar or Campaign Tie-In**
  - Why we're creating this now and how it connects to broader efforts.
- **Voice and Tone Reminders**
  - Key traits from your brand voice chart (e.g. "warm, helpful, direct") that should guide the writing.

Draft Body Structure (Middle Section)

This section is tailored to the content type and helps creators organize ideas clearly.

- **Header or Section Outline**
  - Suggested H2s, bullet prompts, or narrative arc.
- **Key Messages or Talking Points**
  - What must be conveyed, no matter how it's styled.
- **Pull Quotes or Stats**
  - Optional, but great for anchoring points or adding authority.

- **SEO/GAO Elements**

    ◦ Focus keyword(s), question being answered, meta fields.

Editorial Reminders (Bottom Section)

- **Style Considerations**

    ◦ Any format-specific tips (e.g. sentence length, visual support).

- **Accessibility and Compliance Notes**

    ◦ Alt text, captions, disclaimers, or localization flags.

- **Internal Links and CTAs**

    ◦ Reference any related content or assets to include.

Format-Specific Examples

Here are a few examples of how your format or challenge-specific templates can vary:

- **Blog Post Draft Template** – Includes H2/H3 structure, metadata prompts, and CTA placement.
- **LinkedIn Livestream Template** – Focuses on strong hooks, conversational tone, includes all the fields required to schedule the event on the platform, and the CTA of registering to attend or leaving questions in the comments.
- **YouTube Script Template** – Includes intro hook, segment structure, callouts to visuals, closing CTA, and all the input fields required to publish the video.
- **E-book/Guide Draft Template** – Focuses on modular sections, gated content strategy, and CTA integration for lead nurture.

This modular approach lets your team move faster while preserving brand consistency—and it dramatically improves the quality of first drafts. No more guesswork, and fewer rewrites.

Structure Tip

While each template will vary, start with voice and audience intent—always. It's easier (and faster) to write content that hits the mark when creators are reminded up front who they're talking to and why it matters.

SEO, GAO, and GEO Checklist

Ensure content is discoverable, optimized for both search engines and generative engines, and technically sound from the moment it's drafted.

Most content teams are familiar with SEO checklists—such as meta titles, focus keywords, and the like. But today, optimizing for organic visibility requires a broader view. That's where GAO (Generative Answer Optimization) and GEO (Generative Engine Optimization) come in.

These emerging disciplines focus on how your content performs in AI-powered environments, including chatbots, search engine result summaries, voice assistants, and more. In other words, are you showing up when an AI is doing the searching on behalf of your audience?

To prepare for this shift, your templates should embed optimization tasks upfront, during content planning—not tacked on during publishing.

Core SEO + GAO + GEO Fields to Include in All Web Content Templates

- **Meta Title (SEO):** Clear, keyword-aligned, 60 characters max.
- **Meta Description (SEO):** Compelling, accurate summary, under 155 characters.
- **Focus Keyword(s) (SEO/GAO):** Phrase your audience is actually searching or asking about.
- **URL Slug (SEO):** Clean, keyword-relevant, no stop words (articles, conjunctions, prepositions) or unnecessary numbers.
- **Image Alt Text (SEO/Accessibility):** Describes the image and its meaning in context. Resist the urge to shoehorn in keywords.
- **One-Sentence Summary (GAO/GEO):** This becomes the generative engine's go-to descriptor. Think of it as your "TL;DR" for bots.
- **Answer Intent (GAO):** What specific question is this content answering? Phrase it as a question and a one-line response.

- **Entity Alignment (GEO):** Ensure proper nouns (products, people, companies) are mentioned clearly and consistently to reinforce knowledge graph associations.

Structure Tip

Place this checklist in the first third of your content templates—preferably right after the audience and objective section. This ensures optimization is part of the drafting process, not something bolted on post-approval.

Review & QA Checklist

This checklist ensures that every piece of content leaving your team is consistent, polished, and aligned with your brand standards. It systematizes quality control across all formats and contributors—especially important when multiple hands touch the content before it goes live.

It's your last line of defense before publishing. A strong quality assurance (QA) checklist helps avoid embarrassing errors, off-brand moments, and legal or ethical missteps that can erode trust with your audience.

What to Include

This list can be customized by content format, but the following elements form the baseline of any effective QA process:

Voice, Tone, & Style

- Does the content reflect our brand's voice traits (e.g., confident, helpful, conversational)?
- Is the tone appropriate for the audience and format?
- Does it follow our approved style guide (e.g., AP, Chicago, or internal style)?

Strategic Alignment

- Does the content clearly support a content pillar or campaign objective?
- Is the CTA aligned with the intended journey stage?
- Have the intended persona's needs been directly addressed?

Branding & Formatting

- Are logos, product names, and terminology used consistently and correctly?
- Is the formatting accessible (e.g., proper use of headings, lists, alt text)?
- Are brand-approved visuals or templates being used correctly?

Content Accuracy

- Are all statistics, facts, and citations accurate and sourced?
- Are internal and external links functional and relevant?
- Is product or service information current?

Compliance & Code of Conduct

- Are disclaimers or legal language included (if required)?
- Does the piece avoid biased, exclusionary, or potentially problematic phrasing?
- Has the content been checked against accessibility guidelines (e.g., image alt-text, descriptive link text, color contrast)?

Channel-Specific Elements

Tailor checklists for unique publishing needs, such as:

- Blog: Meta title, slug, alt text, author bio
- Social: Platform-appropriate formatting, hashtags, character count
- Email: Pre-header, unsubscribe language, fallback content
- Video/Audio: Closed captions, transcripts, clear audio mix

Structure Tip

Make this checklist easily accessible and fillable—either embedded in your CMS workflow, included as a review tab in your content

tracker, or dropped into your collaborative docs as a final step before approval. If it's not easy to use, it won't be used. This checklist can also be easily converted into an AI workflow to expedite the review process.

This checklist helps standardize review—whether your content is coming from internal teammates, freelancers, or SME contributors. It's also a training tool: the more your team uses it, the more these quality checks become second nature.

Distribution & Promotion Planner

Publishing is just the halfway point for content. This planner ensures that every piece of content gets the post-launch attention it deserves—so your high-effort content doesn't die in quiet corners of your CMS. It brings structure to your distribution workflows by mapping out where, how, and when content will be promoted—and who's responsible for making it happen.

It's one of the most overlooked pieces of content operations. But without this kind of planning, content won't reach its full potential—no matter how well it was written or designed.

What to Include

This template works best as a standalone tab within your content inventory tracker or embedded in your editorial calendar view. You can also mirror it in an existing project management tool. Here's what to capture:

Distribution Channels

Outline both internal and external distribution points, such as:

- Owned Channels: Email newsletter, blog homepage, social channels, in-product messaging
- Paid Channels: Boosted posts, sponsored content, retargeting ads
- Earned/Shared: Partner promotions, influencer sharing, PR placements
- Internal Teams: Sales enablement, customer success, internal newsletters, or Slack

Timing & Cadence

Schedule each promotional push, not just a one-and-done share on launch day:

- Pre-launch teaser (optional)
- Launch day
- Day Three post
- Week Two roundup
- Monthly newsletter inclusion

Evergreen content deserves ongoing promotion—note dates for future reshares or campaign tie-ins.

Roles & Responsibilities

- Who creates the social copy?
- Who builds the email asset?
- Who schedules the posts?
- Who monitors engagement?

Clear handoffs prevent content from stalling in the "someone should share this" stage.

Structure Tip

Tie this planner directly to your content tracker. When a blog post or guide reaches "Final" status, it should automatically flag the need for distribution planning. If your system allows it, trigger a notification to your marketing lead or communications partner to start building the promotion plan immediately.

This planner helps you operationalize content amplification—and make it consistent across formats and teams. Instead of relying on one-off posts or chasing visibility after the fact, you build a habit of proactive promotion that turns every asset into a campaign driver.

Editorial Calendar

Your editorial calendar is your publishing compass. It allows you to zoom out, spot gaps, align around campaigns, and pace your publishing rhythm. It's not just about keeping track of dates—it's how you plan content with intention, get stakeholder visibility, and avoid the "last-minute scramble" that derails strategy.

Used consistently, your calendar becomes the connective tissue between strategy and execution—helping you prioritize what gets published and when, based on what your audience actually needs and what your business is focused on.

What to Include

Whether you use Google Calendar, Notion, Trello, Airtable, or another tool, your calendar should include fields that give both high-level visibility and tactical clarity:

- **Content Title**
  - The working or final title of the piece, linked to the brief or draft when possible.
- **Publish Date**
  - The confirmed or target date the asset will go live.
- **Content Format**
  - Blog post, newsletter, webinar, podcast, infographic, social campaign, etc.
- **Assigned Owner**
  - Who is responsible for delivery and/or publication.
- **Status**
  - Use consistent, color-coded labels like: Idea | In Draft | In Review | Final | Scheduled | Published.
- **Pillar or Theme**
  - Map each piece to a core pillar or campaign to ensure balanced coverage.
- **Campaign or Event Tie-In**
  - Include product launches, seasonal events, industry trends, or partner timelines.
- **Distribution Notes**
  - Optional: Add a brief plan for where and how the piece will be promoted post-launch.

Structure Tip

Choose a shared, collaborative tool and build multiple calendar views (e.g., weekly publishing view, campaign view, persona-specific content view). That flexibility helps everyone—from writers to execs—see what they need without wading through clutter.

If you're using a more advanced system like Airtable or Notion, connect your editorial calendar directly to your content tracker and distribution planner for seamless handoffs and status syncing.

Why You Really Do Need a Content Calendar

The editorial calendar isn't just a content operations artifact—it's a strategic alignment tool. It helps answer key questions like:

- Are we publishing consistently across all key channels?
- Are we clustering too much content around one topic or format?
- Are we creating enough lead time for campaign tie-ins or seasonal content?
- Are we leaving room for reactive content or timely thought leadership?

It also helps manage capacity. By visualizing what's due and when, you can avoid burnout, overcommitment, and last-minute bottlenecks.

When paired with your Idea-to-Execution Tracker and your Distribution Planner, the editorial calendar becomes your team's single source of truth for content visibility, timing, and momentum.

Content Inventory & Audit Tracker

Great content doesn't stop working when it's published—but it does require upkeep. This tracker helps you identify which assets are outdated, off-brand, underperforming, or ready for a strategic refresh. Instead of treating content as a one-and-done deliverable, this template reframes it as a living part of your marketing ecosystem.

When you're working at scale, a systematized audit process is the only way to keep your content library aligned with your current goals, messaging, and audience needs.

What to Track

Whether you're auditing quarterly, twice a year, or tied to campaign planning, your content audit tracker should include the following fields:

- Content Title
- The published title or working name of the asset.
- URL / Location
- Where it lives (internal system, CMS link, gated asset folder, etc.).
- Content Type / Format
- Blog post, landing page, video, guide, webinar, etc.
- Publish Date / Last Updated
- Helps you catch assets that may be outdated or irrelevant.
- Pillar or Topic
- What theme or category it supports—use your pillar framework here.
- Performance Snapshot
- Optional, but powerful: Include key metrics like traffic, engagement, or conversion rate to help prioritize.
- Refresh Needed?
  - Use simple tags like:
    - Yes – Outdated or misaligned
    - Partial – Needs light updates or optimization
    - No – Still performing and current.
- Recommended Action
  - E.g., Retire, Repurpose, Refresh, Reposition, Re-promote.
- Owner + Timeline
  - Who's responsible for the update and when it should be completed.

Structure Tip

Embed this tracker into your quarterly or biannual planning cycle. Don't audit your entire library at once—focus on a specific content type (like top blog posts), channel, or customer stage during each cycle to keep it manageable. For ease of maintenance, make adding new content to this tracker the final step in every publishing workflow.

Whether you use a spreadsheet or a project tool, set up filters so you can sort by age, performance, or update status at a glance.

Tie your refresh planning to your content calendar.

Updated content can be just as impactful (and often faster to publish) than net-new content. When your audit identifies a high-performing post that's two years old, you've just found an easy win for next month's calendar.

Content refreshes are also ideal candidates for:

- SEO + GAO

- CTA testing

- New visuals or format upgrades (e.g., adding video or a downloadable checklist)

A well-maintained content library boosts discoverability, strengthens brand authority, and maximizes the ROI of your existing efforts. This tracker makes that maintenance doable—and repeatable.

This list of templates might initially feel overwhelming to consider. But once they are in place, they bring a much-needed sense of order and consistency to your content efforts.

Here's why these templates work:

- Strategic-first: Each template begins with intent—who you're serving and why—not just what you're creating.

- Process-aligned: They flow from one stage to the next, helping you avoid "start-stop" paralysis.

- Team-friendly: Shareable, accessible, and clear—these templates reduce friction when collaborating with freelancers or in-house talent.

- Efficiency drivers: Templates make content creation less mentally exhausting and more repeatable, freeing you to focus on value—not structure.

Easy-to-Replicate Processes That Reduce Friction and Overthinking

Repeatable processes aren't about taking creativity out of content—they're about taking decision fatigue out of the workflow.

When your processes are documented, accessible, and easy to follow, your content creators don't have to ask:

- Where do I submit this request?
- Who needs to approve this?
- What's the deadline?
- Did we remember the metadata?

Instead, they can focus on what matters: making great content.

Let's look at a few core processes that reduce friction—especially when you're managing multiple contributors, SMEs, or AI tools.

Start with an Intake Process That Sets You Up for Success

When every request comes in via Slack or a hallway chat, you end up with duplicate efforts, unclear goals, and random acts of content.

That's why I always implement a content intake form that captures:

- Content requestor + approver
- Target audience + buyer stage
- Objective + key message
- Deadline + distribution plan

This simple form keeps new content aligned with your strategy before your team spends time writing. It also makes it easier to say "not right now" to requests that don't support your goals—without playing the bad guy.

Document Your Workflows So No One Has to Guess

If your team is constantly wondering who's responsible for what—

or waiting on silent approvals—your process isn't just inefficient, it's stressful.

I map out key workflows for:

- Content intake + idea approval
- Drafting + internal reviews
- Stakeholder feedback
- Final QA + publishing
- Distribution + performance tracking

These can live in a simple Miro board, Google Doc, or even a slide deck. The format doesn't matter—clarity does.

Once you've documented the process, people can jump in confidently and projects don't stall every time someone takes PTO.

Here's a simplified example of a content approval workflow drafted in slide software:

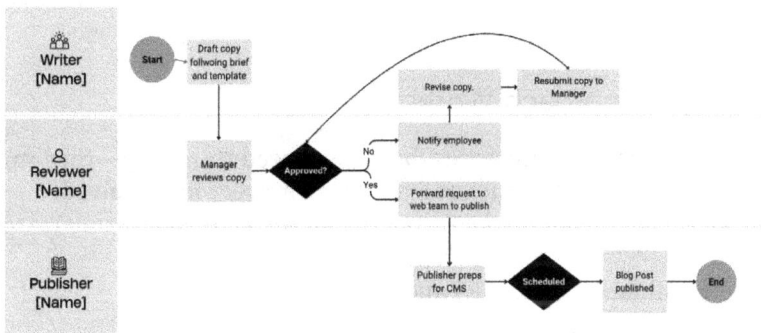

Use Checklists as a Built-In QA Layer

Checklists might seem basic, but they're powerful—especially in fast-paced content environments.

I build them into templates and workflows as quick-reference guides:

- Is this aligned with our brand voice?
- Is the CTA clear?
- Are all required metadata fields filled?
- Have we included internal proof points or data?
- Did the SME review the final draft?

Over time, these checklists become second nature. But having them written down prevents small things from slipping through the cracks— and eliminates back-and-forth at the finish line.

Make It Easy to Reuse and Repurpose Content

When your content follows a consistent structure, repurposing isn't an afterthought—it's built in.

You can easily spin a single asset into:

- Social posts (based on headline angles or CTAs)
- Sales enablement snippets (from your key messages)
- Newsletter blurbs (using the POV summary)
- Slide content (pulled from proof points)

No need to rewrite or reverse engineer. The raw material is already there, formatted and tagged for reuse—and ready to be plugged into your templates.

Why This Approach Works

Easy-to-replicate processes don't just speed things up—they make the entire content experience smoother for everyone involved.

They help you:

- Eliminate bottlenecks and confusion
- Onboard new collaborators faster
- Increase output without increasing burnout
- Maintain quality—even at scale

Templates give you structure. Processes give you momentum. And when both are in place, your content engine runs a whole lot smoother.

Tools That Support Faster Research and Smarter Topic Selection

Even the best content templates can't save you if you're focused on the wrong ideas—or chasing every trending topic with no clear payoff. Before you open your keyword tool or start prompting your favorite GenAI platform, you need to listen first.

What are your ideal customers already asking? Which objections keep surfacing in sales pitches? What are your subject-matter experts tired of repeating? That's your listening research goldmine. These signals give you a far more authentic starting point than any search volume chart—and they're the raw material for smarter, more relevant content.

Start with Signals, Not Search Volume

I don't just look at keywords. I listen. In my own process for clients, I pull topic ideas from customer service queries, social media comments, Slack threads between internal experts, and LinkedIn comments. These real conversations tell you what your audience is struggling with—often long before SEO tools catch up.

For a solo content creator, this might mean pulling recurring questions from your DMs or jotting down pain points that pop up in webinar chats.

In an enterprise setting, it could mean mining sales call transcripts at scale or reviewing help desk tickets across regions. Either way, the goal is the same: tune into the unfiltered language your audience already uses.

That language doesn't just inspire topics—it becomes a powerful input for AI ideation when you feed it into tools as raw, real-world data.

Use AI and SEO Tools as Accelerators—Not Autopilot

AI is an incredible assistant when it's well-trained—and a disaster when it isn't. I use tools like ChatGPT, Perplexity, and Semrush to cluster related ideas, spot content gaps, and brainstorm different angles for specific buyer stages. But AI works best when you treat it like an eager intern: give it context, direction, and your brand voice, or it will guess and hand you back generic fluff.

A solo creator might use AI to quickly brainstorm five variations of a blog post idea pulled from a client's question. A larger team might analyze competitor libraries to surface outdated advice and feed those gaps into their editorial calendar. In both cases, AI isn't the driver—it's an accelerator.

Keep a Living Topic Tracker

Not every idea deserves to become content—but every viable idea should live somewhere visible. A shared tracker gives your team a place to capture signals, see priorities at a glance, and understand why certain ideas are moving forward while others aren't.

For solo creators, this might be a simple spreadsheet of ideas with audience notes and funnel stages. For enterprise teams, it could be a shared project management board that connects content opportunities to broader campaigns. In both cases, the tracker turns ideation from a chaotic free-for-all into a strategic pipeline of future content.

Make Research Reusable

Research is too valuable to vanish into someone's personal Google Drive. By centralizing transcripts, survey insights, and annotated AI prompts, you create compounding value. Each campaign builds on the last, so your second blog post is faster and sharper than your first.

For a solo creator, that might mean building a single master doc of approved stats and go-to references. For a larger team, it might mean maintaining a shared knowledge base or wiki that makes vetted research accessible across regions and departments.

Why This Works

Used the right way, research tools don't just speed things up—they sharpen your strategy. You end up with content built on audience

needs, not assumptions. You prioritize ideas that actually move the needle. You make ideation collaborative instead of chaotic. And most importantly, you spend less time reinventing the wheel and more time creating content that matters.

Systems That Help You Prioritize Ideas That Actually Align with Your Goals

You'll never run out of content ideas. But you'll always run out of time and capacity.

That's why idea generation isn't the hard part—it's idea prioritization.

Without a system, you end up chasing whatever's loudest: a stakeholder's last-minute ask, a trending meme, or a brilliant idea with no clear business case. That's how content turns into clutter.

Let's walk through how to build prioritization systems (for solo creators and large teams alike) that act like strategic filters—so your best ideas rise to the top and support your content goals.

Start with a Definition of What "Strategic" Means to You

Before you can rank ideas, you need to know what matters.

For me, a strategic content idea is one that's aligned with at least one of these:

- A targeted marketing campaign or business priority
- A core audience persona and journey stage
- A relevant product or service line
- A clear, measurable objective (e.g., lead generation, education, engagement)

I include these fields in every content intake form and idea tracker.

That way, we're not just collecting ideas—we're collecting context.

Use a Simple Scoring System to Rank What Matters Most

When I was the owner of the B2B email channel for a large company with many competing priorities, we could have easily bombarded our customers with an email every day of the week if we had said yes to every request that came in.

But luckily, our leaders were aligned on the idea of not sending any specific customer more than a handful of emails per month. With that backing, I worked with my analytics team to devise a simple scoring system to rank email performance. Once that was well-received, we created a version that allowed us to use that performance data as a prioritization input to decide which messages went out to each customer.

Today, you don't need such a heavy lift or any fancy tools to create a consistent prioritization method. For example, you can add some scoring columns to your idea tracker:

- Audience relevance (Is this something they care about?)
- Strategic alignment (Does it support a defined goal or initiative?)
- Differentiation (Can we say something new—or better—than our competitors?)
- Effort vs. impact (Do we have the resources to do this well?)
- Timeliness (Is now the right moment for this piece?)

Each idea gets a 1–5 score per category. The total gives you a rough "priority index"—but the discussion this ranking sparks is the real value. You still may end up with some low-scoring ideas becoming a top priority, but everyone involved in your process will be much more aware of the tradeoff they're making.

Track Decisions (and Revisit Them)

Prioritization isn't a one-time task. It's an ongoing conversation. One way to make sure that happens is to embed prioritization into your content planning tools or worksheets:

- "Why now" notes (for approved content)
- "Hold for later" sections (for good ideas with unclear timing)
- Performance fields (to track which bets paid off)

This builds a learning loop so your future prioritization isn't just gut feel. It's informed by real outcomes.

Why Prioritization Works

A good prioritization system doesn't kill creativity. It protects it. When your team knows how ideas are evaluated, they stop chasing approval and start thinking more strategically. You reduce content chaos and random pivots, making better use of your team's time and energy. You can then focus on fewer, higher-impact pieces, which are much more likely to drive engagement.

When everything feels important, nothing gets done. But when you know what matters—and have a system to stay focused—you create content that moves your business forward.

Align with Content Stakeholders Early—So You Don't Have to Backpedal Later

Prioritization shouldn't happen in a vacuum.

If your stakeholders feel blindsided when their idea gets bumped, your system will break. That's why I include prioritization checkpoints in:

- Editorial board meetings (monthly or quarterly)
- Campaign planning sessions
- Stakeholder review loops (built into our templates)

When everyone sees the same roadmap, it's easier to explain why something is a "yes later," not a "no forever." And by doing so, you create alignment across departments—without having to "sell" every decision.

How to Build a Sustainable Content Planning Workflow

Sustainability in content planning doesn't just result in less burnout or fewer missed deadlines. It means your process can scale—with clarity, consistency, and confidence.

A sustainable content workflow makes it easier to stick to strategic priorities, set realistic expectations, and adjust when things inevitably shift. In essence, it allows you to collaborate without chaos. Let's walk through the building blocks of a workflow that can support high-quality content—month after month, quarter after quarter.

Start with a Shared Calendar—but Don't Live by It

Your editorial calendar is the most visible part of your workflow, but it's not the whole thing. I treat the calendar as a reflection of three planning layers:

1. Strategic campaigns (quarterly or annual focus areas)

2. Content themes (monthly or topic-based)

3. Individual pieces (assigned to owners, formats, and due dates)

This tiered approach prevents calendar chaos and helps your team understand not just what you're publishing—but why. Build your calendar around publishing cadence, not just content types. This makes it easier to budget time, plan resources, and spot bottlenecks early.

Work Backward from Publish Dates (and Buffer Generously)

To avoid last-minute scrambles, build a planning timeline that works in reverse. For each content type, define:

- Minimum time needed for research + outlining
- Drafting + review rounds
- Stakeholder or SME approvals
- Final edits, metadata, and CMS prep
- Distribution prep (internal, social, newsletter, etc.)

Add buffer time between each step—and communicate that buffer upfront. This helps protect your deadlines and your relationships.

Use Templates and Checklists to Stay Consistent

A sustainable workflow shouldn't rely on your memory. I include key workflow steps in my content templates and project trackers:

- Intake form submitted + approved?
- SME or customer quotes gathered?
- Internal review completed?
- SEO + metadata added?
- Publish + distribution confirmed?

These aren't just QA tools—they're capacity savers. They free up your team's brainpower to focus on craft, not logistics.

Build in Flexibility for Real Life

Even the most airtight workflow needs room for shifting priorities. That's why I recommend leaving 20–30% of your calendar as "flex slots." These can be used for:

- Timely content tied to events or industry news
- Internal requests that meet strategic criteria
- Fast-turn content repurposing

By planning for the unexpected, you're less likely to blow up your process every time something changes.

Communicate the Workflow—and Stick to It

A workflow only works if people respect it. I make sure all collaborators—freelancers, SMEs, exec sponsors—know what to expect:

- When content needs to be ready
- What format it should be in
- How and when feedback is gathered
- Who owns final approval

Then I reinforce the process by linking to workflow docs from every intake form, template, and editorial meeting agenda.

How the Process Pays Off

A sustainable content planning workflow turns good intentions into repeatable execution.

It helps your team:

- Avoid deadline-driven panic
- Publish consistently without burnout
- Collaborate more smoothly across roles
- Maintain quality—even as volume increases

Because sustainable doesn't mean slow or boring. It means you have the infrastructure to move fast without breaking things.

Yes, Solo Creators Need to Document Process Too!

It may feel weird to start creating trackers and documenting content workflows when you are a team of one. I get it! But the thing is, if you don't document it, you can't delegate it. Which means you can't scale your content.

This is the upfront work that will allow you to focus on the work you love and outsource the rest while retaining what makes your content unique.

Example Workflows for Sustainable Content Planning

A workflow isn't just a checklist—it's a shared agreement. For a workflow to be sustainable, every step needs a clear owner. That means defining:

- Who's responsible for moving each step forward
- Who's reviewing or approving the work
- What tools or templates support that step

Without ownership, even the best workflows fall apart. Tasks stall in inboxes, approvals get delayed, and content gets rushed at the finish line.

The following examples show how I have structured real-life content workflows—from idea intake to publication and repurposing. They're designed to keep teams aligned and projects moving—without relying on a single person to carry the process alone.

These workflows reflect how I structure repeatable content processes for internal teams, freelancers, and SME collaborators. They're not rigid checklists—they're scaffolding to keep your team aligned, efficient, and sane.

1. Blog Post Workflow (Collaborative / SME-Driven)

Week 1: Idea Intake + Alignment

- Content request submitted via form
- Content strategist reviews for strategic alignment
- Topic added to editorial calendar and tracker
- Assigned to the writer with a draft due date and SME contact

Week 2: Drafting + SME Input

- Brief created and shared with the writer
- SME quotes/interview (if needed)
- Draft v1 submitted for internal review

Week 3: Review + Finalization

- Review by content strategist (clarity, brand voice, SEO)
- Feedback sent to the writer
- Final draft submitted with:
  - Meta title + description
  - SEO keyword
  - Internal links + CTA
- Final approval from SME

Week 4: Publish + Distribute

- Post scheduled in CMS with correct formatting
- Shared in newsletter, LinkedIn, and sales Slack channel
- Added to internal asset tracker for repurposing

2. Content Request to Publish (for Campaign Content or Thought Leadership)

Step 1: Submission
- Stakeholder completes the request form with:
    - Objective
    - Audience
    - Key message
    - CTA
    - Deadline
    - Distribution plan

Step 2: Review + Kickoff
- Weekly content planning meeting: review all new requests
- Approve, refine, or suggest alternate formats
- Assign team lead (writer or PM) and define deliverables

Step 3: Production
- Brief created
- Draft submitted + edited
- Approvals gathered (SME, marketing, legal, if needed)

Step 4: QA + Publication
- Final copy added to design or CMS
- Internal checklist completed (links, voice, accessibility)
- Distributed per request plan

3. Repurposing Workflow (From Webinar to Written Content)

Event Complete
- Transcription pulled via AI tool (Otter, Descript, etc.)
- Content strategist reviews themes and notes key timestamps

Repurposing Breakdown
- Blog post with 2–3 key takeaways
- Quote cards or stats for LinkedIn

- Short-form video or audiogram clips
- Newsletter summary (with CTA to full replay)

Assign + Track
- Assign asset owners + due dates
- Add to editorial calendar as "Repurpose" assets
- Link all outputs back to the original event for performance tracking

4. Monthly Editorial Cycle (Sustainable Team Rhythm)

Week 1: Planning
- Review performance metrics from the previous month
- Approve upcoming ideas
- Assign tasks + confirm SMEs and deadlines

Week 2–3: Creation + Review
- Writers draft + revise
- SMEs provide input
- Review cycles completed

Week 4: Finalization + Reporting
- Content scheduled, published, and distributed
- Update tracker with final links and publish dates
- Monthly content wrap-up report shared with stakeholders

Why These Workflows Work

Documented workflows aren't about creating complexity—they're about providing clarity. Each one:
- Starts with alignment (via request or planning)
- Breaks down roles and expectations
- Bakes in time for feedback + review
- Ends with publication and reuse

They're repeatable, teachable, and adaptable as your content program grows.

CHAPTER 6

Collaborating with External and Internal Talent

No matter how experienced or talented you are, content creation is a team sport.

From solo startups outsourcing their first blog post to enterprise content teams managing a hybrid mix of agencies, freelancers, and internal collaborators—one constant remains: successful content programs rely on people working together, not just in parallel.

But collaboration isn't automatic. It takes more than a shared Google Doc and a few Slack messages to get it right.

In this chapter, we'll explore how to build the infrastructure that makes collaboration effective, efficient, and sustainable—whether you're working with freelance writers, ghostwriters, designers, developers, or your future in-house team. That includes:

- Working effectively with freelancers, agencies, and cross-functional partners
- Creating onboarding documentation that sets everyone up for success
- Using project management tools to reduce chaos and increase visibility
- Setting clear expectations for quality, timing, and communication

We'll explore the tools that will allow you to collaborate with purpose—and avoid the bottlenecks and burnout that can sink even the most promising content strategies.

Working Effectively with Freelancers, Agencies, and Cross-Functional Partners

Freelancers, agencies, and internal partners are the engines behind scalable content programs—but only when clarity and collaboration are built into the framework. Without deliberate structure, projects can bog down, misalign, or stall completely. To get—and keep—everyone rowing in the same direction, you need purpose-built systems for alignment, accountability, and communication.

1. Set Clear Expectations at Onboarding

Every contributor starts the same way: wanting to do good work. But good intentions can falter without clarity around your brand, process, and timeline.

An onboarding doc helps answer foundational questions:

- What's our brand voice and value proposition?
- Who's the audience and what's their intent?
- What's the expected deliverable format and deadline?
- Who's reviewing and who's approving?

This isn't micromanagement—it's a gift of context. One thoughtful freelancer might not fall into defaulting on assumptions, but a brief that spells out your room for creativity, feedback rhythm, and deliverables lets them excel with confidence.

2. Align Stakeholders and External Contributors

Internal silos are often where friction starts. When content, PR, and social teams operate out of sync, work gets duplicated—or worse, derailed. Break the silos by:

- Inviting all relevant contributors into planning sessions—not just for visibility, but for shared ownership

- Stating shared goals clearly ("This blog post supports a lead-gen campaign; that press release is building top-of-funnel awareness")
- Agreeing on workflows: who drafts, who reviews, who schedules

Using shared goals and timelines keeps cross-functional collaborators aligned and content moving forward instead of bouncing back.

3. Define Roles and Communication Rhythms

Let's be clear: Communication isn't optional. Each contributor needs to know:

- Who they report to
- When updates are due
- Which tools to use (project manager, email, shared docs)

From my experience with distributed teams, having these three frameworks in place is the difference between a smooth content marketing program and a chaotic mess. Naming the owner explicitly for drafts, feedback, approvals, and publishing keeps the work flowing through the system.

Creating Onboarding Documentation That Sets Everyone up for Success

A strong onboarding experience doesn't just introduce your freelancers or new hires to your brand—it empowers them to contribute confidently from day one.

When you treat onboarding as a strategic content asset, not a last-minute email or Slack thread, you eliminate confusion, prevent bottlenecks, and raise the bar for quality across every contributor.

Think of it as your first chance to scale trust.

Why Onboarding Documentation Matters

Without onboarding documentation, your new team members are left to guess:

- What "good" looks like for your content
- How to reflect your brand's voice and tone

- Who to go to with questions or blockers
- What tools and workflows they're expected to use

That ambiguity leads to rework, frustration, and missed deadlines—especially for remote or freelance collaborators who aren't looped into every internal nuance.

As content strategist Marcia Trask noted in her #ContentChat episode: "You can't assume your collaborators know what success looks like unless you show them—explicitly.[10]"

What to Include in Your Onboarding Packet

A solid onboarding doc or wiki page doesn't need to be fancy—it just needs to be complete and accessible. At minimum, include:

Brand and Audience Overview

- Brand voice guidelines + tone examples
- Audience personas or ICP snapshots
- Company elevator pitch or messaging hierarchy

Content Expectations

- Sample content that reflects your ideal quality
- Content types you publish (and don't)
- Common formatting, length, or structure notes

Processes and Tools

- Content brief template or intake form
- Project management platform + login info
- Where drafts are submitted + how feedback is delivered

10 Trask, Marcia. "#ContentChat Measuring Content Marketing's Results." Erika Heald Marketing Consulting. December 2, 2024. https://erikaheald. com/content-chat-how-to-measure-content-marketing-results/. Accessed 21 October 2025.

Roles and Approvals

- Who reviews and approves content
- Typical turnaround time expectations
- Who to contact with questions or blockers

Style and SEO Guidelines

- Link to your editorial or content style guide
- SEO requirements (e.g., metadata, keyword placement, alt text)
- Accessibility considerations or tone inclusivity rules

Make It a Living Document

Your onboarding docs shouldn't sit untouched for a year. Treat them like a content asset:

- Review quarterly (or whenever workflows change)
- Invite feedback from your team and freelancers
- Use version history or comments to clarify gray areas

A simple Google Doc, Notion page, or shared wiki is enough—as long as it's regularly updated and easy to find.

Why This Works

Clear onboarding documentation saves time on both ends. It reduces back-and-forth questions, speeds up your content pipeline, and gives contributors the confidence to deliver their best work—faster.

More importantly, it sends a message: "We value your time, and we're setting you up to succeed."

When you extend that kind of clarity and care, your content reflects it.

Creating an Editorial & Contributor Guide

Your editorial or contributor guide is more than a style sheet—it's your onboarding tool for great content. It gives internal teams, freelancers, SMEs, and even AI collaborators the context they need to contribute effectively, consistently, and on-brand.

Without it, every piece becomes a custom coaching session. With it, you empower contributors to hit the ground running—minimizing back-and-forth, rewrites, and frustration on all sides.

This guide is the glue between your strategy and your execution—especially when multiple voices are involved.

What to Include

The exact structure can vary by team size or contributor type, but every solid contributor guide should cover the following areas:

- Audience Overview

  - Quick reference on personas, journey stages, and key challenges
  - What each audience cares about, and how your content supports them

- Voice & Tone Guidelines

  - Define your brand's voice traits (e.g., confident, curious, warm)
  - Offer before/after examples or "say this/not that" lists
  - Clarify tone shifts by channel or content type (e.g., more casual on LinkedIn)

- Content Expectations

  - What "good" looks like for your organization
  - Preferred content structure or length by format (e.g., blog posts vs. case studies)
  - Whether AI-generated content is allowed (and if so, how it must be fact-checked or edited)

- Process & Workflow

  - How and where to submit drafts (Google Docs? CMS?)
  - Review stages and who's involved in approvals
  - Typical turnaround time and version control tips

- SEO / GAO / Accessibility Standards

  ◦ Keyword and metadata requirements

  ◦ Link guidelines (internal vs. external)

  ◦ Accessibility considerations (alt text, heading hierarchy, inclusive language)

- Legal, Ethical, and Compliance Notes

  ◦ Citation requirements for quotes and data

  ◦ Disclosures for affiliate or partner content

  ◦ Policies for referencing customers, competitors, or sensitive topics

Structure Tip

Make your guidelines modular and skimmable. Use headers, callouts, and linked sections so contributors can find what they need fast—especially freelancers or SMEs who are juggling multiple projects.

Store it somewhere accessible (Notion, Google Drive, CMS, wiki) and link to it in every content brief or intake form. The more frictionlessly you can surface it, the more consistently it'll be followed.

Collaborating with AI? Include This Too:

If you're using AI as a co-creator, this guide helps set the tone and boundaries for prompts and editing. Consider adding:

- Approved voice/tone prompts
- Sample "starter" prompts for each format
- Reminders about fact-checking, citation, and brand-sensitive language
- What AI can help with (e.g., draft outlines) vs. what should stay human-led (e.g., POV pieces)

When everyone has the same frameworks your content process gets faster, cleaner, and more consistent—whether your collaborators are in-house, external, or AI-assisted.

Using Project Management Tools to Reduce Chaos and Increase Visibility

Even the most talented content team can crumble under the weight of email threads, forgotten deadlines, and too many spreadsheets.

That's where project management tools come in—not just to track tasks, but to bring structure, accountability, and calm to your entire content operation.

Whether you're working with freelancers across time zones or juggling priorities with an in-house team, a shared platform becomes your source of truth. It tells you:

- What's being worked on
- Who owns it
- Where it's stuck
- When it's due
- How it ladders up to bigger goals

It replaces guesswork with clarity—and lets everyone focus on the work instead of the workflow.

What a Good Project Management Tool Should Do

You don't need the flashiest platform. You need a tool that:

- Makes deadlines visible
- Assigns and tracks ownership
- Houses files and feedback in one place
- Plays well with your other tools (Docs, Slack, email)
- Is easy enough that people actually use it

Whether you're building a content calendar, managing campaign deliverables, or tracking review cycles, the goal is to make your process visible and repeatable.

Project Tools That Work for Content Teams

Here are a few tools I've used or seen work well across different team sizes and content types:

- **Asana**

  - Flexible task management with timeline and calendar views, perfect for mapping out content pipelines and campaigns. You can create content templates, assign collaborators, and visualize bandwidth at a glance.

- **Redbooth**

  - Simple and streamlined, ideal for smaller teams or agencies managing multiple clients. Its discussion and file-attachment features keep conversations tied to specific tasks. (Disclosure: they are a former client and I use the tool to this day.)

- **Notion**

  - Great for teams that want an all-in-one wiki + planner. Build custom dashboards for content calendars, briefs, SOPs, and performance metrics. It's highly customizable—but best with someone to maintain it.

- **Wrike**

  - Best for enterprise or complex content workflows with layered approvals. Useful when you need dependencies, Gantt charts, or cross-department collaboration at scale.

- **Microsoft Teams (with Planner or To Do)**

  - If your organization already uses Microsoft 365, Teams can become your content HQ. Use integrated task boards, file sharing, and chat threads to centralize collaboration— without adding another tool to your stack.

How to Make the Tool Actually Work for You

Tools don't fix process problems. People do. Here's how to set yours up for success:

- Create shared workflows. Everyone should know how a blog post or campaign moves from idea to publication.

- Assign owners at every step. Don't just assign the task—assign the review, the follow-up, the final check.
- Build templates. Standardize recurring content types so no one's starting from scratch.
- Limit tool sprawl. Stick to one main platform, and integrate it where possible (Slack, Google Drive, calendar).

Why This Works

A good tool doesn't just keep you organized—it reduces stress, improves communication, and protects creative time.

When your team can see what's happening, what's coming, and what's expected of them, they work more confidently and collaboratively.

And that's how great content gets created—on time, on brand, and without the burnout.

Content Planning + Production Workflow in Notion

Here's an example of how you can use Notion as your content collaboration and governance single source of truth.

Sample Content Workflow in Notion

Plan, track, and publish with visibility from idea to live.

Main Dashboard: Content HQ

A high-level, always-updated homepage linking to:

- Editorial Calendar (Database view)
- Content Briefs
- Approved Topics Bank
- Brand + Voice Guidelines
- SME + Reviewer Directory
- Repurposing Tracker
- Performance Snapshots

This page acts as an overview of your content program.

Editorial Calendar (Database View)

Table or calendar view with key fields:

- Title
- Status
- Owner
- Publish Date
- Journey Stage
- Format
- Pillar
- CTA
- Priority

Status Options:

- Idea
- Briefing
- Drafting
- In Review
- Approved
- Scheduled
- Published
- Repurpose Queue

Use Notion filters to create personalized views for writers, editors, and stakeholders.

Content Brief Template (Linked in each calendar entry)

Each content piece links to its own brief (discussed in detail in Chapter 5) with a baseline of these structured fields:

- Audience Persona
- Goal / Funnel Stage
- Core Message or POV
- SME(s) to Interview
- Voice + Tone Guidance

- Internal Links + Sources
- Metadata Checklist (SEO title, meta description, keyword)
- Distribution Notes

Repurposing Tracker

Database or linked section inside the original content brief:

- Which formats it has been spun into (e.g., blog > LinkedIn > infographic)
- Who's owning each format
- Distribution links + performance snapshot

Workflow Automations + Best Practices

- Use templates to pre-populate new briefs
- Set reminders for "Due this week" status
- Link to Google Docs for long-form drafts
- @mention SMEs and reviewers inside each brief for accountability
- Archive completed content monthly—but keep the data for performance analysis

Why This Approach Works

With a project management tool like Notion, your content workflow becomes:

- Centralized: Everything lives in one place (no more buried emails or lost Slack threads)
- Customizable: Tailor views by role, stage, or campaign
- Repeatable: Use templates to remove guesswork and scale faster
- Transparent: Everyone can see the status, expectations, and progress

Setting Clear Expectations for Quality, Timing, and Communication

When content projects stall, miss the mark, or require multiple rounds of revisions, it's rarely due to lack of talent.

More often, the culprit is unclear expectations.

Setting expectations isn't about control. It's about creating clarity so your team, whether internal or external, can do their best work with confidence. And when you're working with freelancers, agencies, or cross-functional collaborators, that clarity becomes even more essential.

What to Define Upfront

To build a shared understanding of success, spell out these three things early:

1. **Quality Standards**

    - What does "good" look like for this brand?
    - Are there content samples or benchmarks to reference?
    - What tone, structure, and level of depth are expected?

    Providing examples and style guides is a shortcut to better output—and less back-and-forth.

2. **Timelines and Turnaround Expectations**

    - What are the target publish dates—and when do drafts/reviews need to happen?
    - How long do reviewers have to provide feedback?
    - What happens when someone's late?

    Include these timeframes in your brief, your project tracker, and your kickoff conversation. Don't leave delivery dates up for interpretation.

3. **Communication Preferences**

    - Which tools will you use to collaborate (email, Slack, Asana)?
    - Who is the primary point of contact?
    - How often should contributors check in?

These seem obvious—until a project slips through the cracks because "I didn't know who to follow up with."

Why This Approach Works

When expectations are clear:

- Contributors feel confident and empowered
- You reduce rework, delays, and last-minute pivots
- Stakeholders are more likely to be aligned and responsive

Most importantly, it creates space for trust—because everyone knows what's expected and how to deliver it.

Clarity isn't just a nice-to-have. It's what makes sustainable collaboration possible.

CHAPTER 7

Reviewing and Polishing Content

Publishing high-quality content isn't just about hitting deadlines. It's about slowing down—at the right moment—to ensure what you're about to ship actually reflects your strategy, your voice, and your standards.

And yet, editing is one of the most overlooked stages in the content lifecycle.

When teams are under pressure to "just get it out," review steps get skipped or rushed. The result? Off-brand messaging, inconsistent tone, SEO gaps, and missed opportunities to resonate with the audience.

This chapter is your reminder—and your roadmap—for why and how to build a content review system that scales with you.

You'll learn how to:

- Define what "quality" means for your team (beyond grammar and spelling)
- Build a repeatable, role-aware content review framework
- Use AI tools to support—not replace—your editing process
- Ensure your content sounds human, aligns with your brand, and connects with your audience

Because whether content is drafted by a freelancer, an in-house SME, or a generative AI tool, the final polish still needs a human touch. That's not old-fashioned—it's how you protect your voice, your credibility, and your results.

Define What "Quality" Means for Your Team (Beyond Grammar and Spelling)

Every marketing team says they want to create and publish "high-quality" content. But without a shared definition of what that actually means, quality becomes subjective—and nearly impossible to replicate at scale.

Quality isn't just fixing typos or polishing punctuation. Truly high-quality content reflects your brand voice and values, supports a clear strategic goal, respects your audience's time and level of expertise, and inspires trust through clarity, originality, and relevance. And, especially in AI-assisted workflows, it must sound human and feel intentional.

Move Beyond "Technically Correct"

Grammatically sound but emotionally flat content doesn't drive conversions—or conversations. That's why your team's definition of quality should be rooted in the experience the content delivers. Ask yourself: Does this sound like us? Does it show empathy for the reader's problem or question? Does it provide a clear, differentiated point of view?

And does it support the goal of the campaign, funnel stage, or business unit? Just because your grammar tool of choice says a turn of phrase is correct doesn't mean it's the right one for your voice.

Make Quality Measurable and Repeatable

To remove ambiguity, document your team's version of "quality" in your editorial guidelines. That definition should include voice and tone pillars (with do/don't examples), clarity around required points of view or takeaways, and minimum standards for originality and value. It should also point to the metrics that signal success—like time on page, shares, or conversions—not just word count.

When everyone—from freelancers to AI prompt writers to final editors—works from the same definition of quality, review cycles shrink and output improves. More importantly, when your content consistently reflects your best thinking, your audience notices. That's quality that builds trust—and results.

Build a Repeatable, Role-Aware Content Review Framework

If your content review process changes with every project—or every contributor—you don't have a system. You have a guessing game.

A strong content review framework ensures that no matter who's involved—freelancer, SME, internal reviewer, or even an AI tool—every piece of content goes through the same checkpoints, in the right order, with the right people accountable for each step. The result? Less confusion, faster approvals, and more consistent quality.

Why Role Clarity Matters

Most review bottlenecks come down to uncertainty. Who owns the first review? Who gives the final green light? Who's just consulted—not a decision-maker? Without clear answers, deadlines slip and quality suffers.

Assigning roles at each stage removes the guesswork:

- Writer: Delivers the initial draft in brand voice, guided by the brief
- First Reviewer: Edits for clarity, structure, and alignment with strategy
- SME or Stakeholder: Confirms technical accuracy and messaging alignment
- Final Editor: Polishes grammar, tone, SEO, and accessibility
- Approver: Provides the final sign-off for publication

Keep the Framework Consistent—but Flexible

Not every piece of content needs the same level of scrutiny. A short social post may only require one set of eyes, while a white paper or product announcement might demand multiple reviewers. The key is to standardize how you decide what's required. Create a checklist by content type, build review workflows into your project management tool, and note review responsibilities directly in your content briefs. That way, no one is left wondering, "Am I supposed to weigh in?"

Build It Into the Brief

Your brief is the best place to set expectations upfront. Clearly outline who's reviewing, what kind of feedback they're responsible for, and when it's due. This avoids crossed wires, unnecessary edits, and the frustration of mismatched expectations. It also aligns everyone around what success looks like before a draft ever hits their inbox.

Why This Works

When roles are clear, collaboration speeds up. When the process is consistent, quality improves. And when every contributor knows what's expected, they're free to focus on what they do best. A review framework doesn't stifle creativity—it protects it.

Use AI Tools to Support—Not Replace— Your Editing Process

AI editing tools can help content teams move faster—but speed only matters if it preserves, or improves, quality. When used thoughtfully, AI becomes a helpful assistant: catching small errors, suggesting smoother phrasing, and surfacing inconsistencies that might otherwise slip through. Without human oversight, however, those same tools can flatten your voice, strip out nuance, or reinforce generic patterns. AI should never be the final reviewer. It should act as your first-pass assistant—one step in a broader, human-centered editing process.

How AI Can Help

AI tools are particularly effective at handling repetitive or mechanical editing tasks. For grammar and readability checks, platforms like Grammarly, Hemingway, or Writer can flag confusing phrasing, identify passive voice or filler language, and highlight inconsistencies in spelling or punctuation. Some even suggest reading-level adjustments tailored to your target audience.

More advanced platforms, such as Writer or Jasper, go further by aligning content with your brand's voice. They can compare drafts against your voice guide, recommend word swaps or tone adjustments, and flag content that feels too formal, too casual, or overly robotic.

AI summarizers and rewriters also offer value when it comes to testing clarity and point of view. They can confirm whether your key message is coming through, demonstrate how you could condense a long-form blog post into a newsletter introduction, or show how variations of the same piece resonate differently across channels.

Guardrails That Protect Your Voice

To make AI helpful—not harmful—you'll need guardrails. Establish clear guidelines for when to use AI and when not to, which tools are approved for editing, and who is responsible for reviewing AI-suggested changes before they are accepted. Most importantly, train your team to approach every AI suggestion with critical thinking and analysis. Just because a tool flags something doesn't mean you should change it!

The Value of Using AI Tools

When integrated responsibly, AI tools help your team catch errors more quickly, refine their work with greater confidence, and reduce fatigue from repetitive, low-level edits. What they cannot replace is the human ability to interpret brand voice, build empathy for the reader, and align every piece of content with a strategic goal. AI accelerates the process, but judgment—and quality—remain human work.

The Importance of Human-First Content

No matter how optimized or well-researched a piece of content may be, it fails if it doesn't feel authentic. Great content is more than clean grammar or clever turns of phrase. It carries a voice your audience instantly recognizes as yours. It's relevant to the challenges they are actually facing.

And it demonstrates empathy—showing that you understand where they are and how you can help.

This matters more than ever as our inboxes and social feeds are overrun by AI-generated copy. If your content reads like it was written for an algorithm instead of a human, people will scroll past—even if the SEO is perfect.

As behavioral marketing expert Nancy Harhut got across in her Content Chat conversation[11], using behavioral psychology to truly connect with

---

11   Irvin, Alek. "November. 4, 2024 #ContentChat Recap: Using Human Behavior and Psychology to Supercharge Your Content Marketing". December 15, 2024. https://erikaheald.com/content-chat-behavioral-psychology-techniques-for-content-marketing/. Accessed 21 October 2025.

your audience through your content means going beyond facts. It's about speaking to how people feel and think, and helping them see themselves in the story you're telling.

Human Content Is Brand Content

What makes your brand stand out isn't just what you say—it's how you say it. Your voice should be unmistakable, whether you're publishing a blog post, sending an email nurture, or writing a quick social caption.

One way to protect that voice is by building "humanity checkpoints" into your review process. Before approving content, ask: Does this sound like something our brand would say? Would our audience feel seen, heard, or helped by it? Does it communicate a clear point of view—or could anyone have written it? These questions keep your content grounded in the personality and perspective that make it uniquely yours.

Connect, Don't Just Inform

Content that connects is content that converts. That's why your final review loop should always include a gut-check for emotional resonance, narrative flow, and relevance to your audience's current pain points or desires. These checks ensure you're not just delivering information but also forging a connection. In an environment where sameness is only a click away, this is what builds trust and loyalty.

A polished piece of content isn't just error-free. It's alive. It has a point of view, a personality, and a purpose. It sounds like you. It serves them. And that stands out—especially in an AI-saturated feed.

PART III

Nurturing Long-Term Growth and Adaptability

The most successful content entrepreneurs don't just create great content They create infrastructure—the behind-the-scenes systems, documents, and habits that keep the business running smoothly as it grows.

This section is about building that infrastructure with intention, because growth brings complexity.

More content. More clients or campaigns. More collaborators. Without a strong foundation, that complexity becomes chaos.

But when your processes are documented, your systems are clear, and your use of AI is strategic—not reactive—you can scale sustainably and adapt confidently.

In the chapters ahead, you'll learn how to:

- Keep your documentation sharp and useful as your business evolves
- Use AI tools to enhance your efficiency without losing your voice
- Strategically grow your team with the right mix of talent and systems

Whether you're a solo creator ready to expand or a small team laying the groundwork for bigger things, this part will help you build a content operation that doesn't just survive growth—but thrives because of it.

CHAPTER 8

Sharpening Documentation and Processes

When your content program starts to grow, it's easy to outpace your original systems.

What once felt streamlined and intuitive can quickly become a bottleneck—especially when you're managing a growing team, juggling new tools, or layering in AI.

That's why documentation isn't a static asset—it's a living part of your business infrastructure. If you treat it like a "set it and forget it" file, you'll find yourself relying on outdated processes that no longer serve your goals.

The teams that scale sustainably are the ones that make documentation and workflow refinement a regular habit. They build a learning culture where systems aren't sacred—they're revisited, tested, and improved over time.

In this chapter, you'll learn how to:

- Create systems for reviewing and refining documentation regularly
- Involve your team in process improvements to build buy-in and efficiency
- Use your documentation to support AI integration and new team member onboarding
- Keep your content operations aligned with your evolving business strategy

Your processes don't need to be perfect—but they do need to be maintained. Continuous refinement is what turns "good enough" into a system that actually supports your growth.

Create Systems for Reviewing and Refining Documentation Regularly

Your documentation is only as useful as it is current. Most teams don't realize their processes are broken until something falls through the cracks—a piece of content goes out with the wrong CTA, a freelance writer misses the tone, or no one knows who is responsible for approvals.

By the time the problem surfaces, trust and time have already been lost.

The key is to build in a rhythm for regularly reviewing and updating your documentation.

This isn't about achieving perfection or creating excessive process. It's about making space to ask important questions: Does this still reflect how we actually work? Are the tools, timelines, and owners still accurate? Have new steps or expectations crept in that aren't documented anywhere? How you answer these questions determines if it's time for a refresh.

Set a Review Cadence—and Stick to It

Treat your documentation like a product with its own release cycle. For fast-growing teams or those frequently adopting new tools, quarterly reviews keep things aligned. Stable teams with well-established workflows may only need biannual updates. And after major milestones— such as completing a content audit or restructuring the team—reviews should happen immediately to ensure the documentation matches the new reality.

One practical way to make this sustainable is to tie documentation reviews to your existing content calendar planning cycles. This ensures process updates happen before a new campaign begins, not in the middle of execution.

Make Documentation Ownership Clear

Every document needs an owner. That doesn't mean one person has to do all the work; it means someone is accountable for keeping it current and flagging when updates are needed. Assign owners to style guides, editorial calendars, content briefs and templates, review workflows, and onboarding materials. Then formalize this responsibility by making documentation review a recurring task in your project management tool.

Invite Feedback from Everyday Content Creators

The best updates come from the people using your documentation every day. Build in checkpoints—through surveys, debriefs, or even quick Slack threads—to ask what feels unclear or outdated, where contributors tend to get stuck, and what should be added or removed to reflect how work actually gets done.

Documentation is not just a leadership task; it's a team sport. When everyone feels empowered to contribute, your systems become more accurate, more useful, and more trusted.

A Documentation Review Checklist

Use this quick audit to keep your systems current and useful.

Step 1: Confirm Accuracy

- Does this document still reflect how we actually work?
- Are tools, timelines, and owners still correct?
- Have new steps or expectations been added elsewhere but not captured here?

Step 2: Check for Relevance

- Is this information actively used by the team?
- Does it support current campaigns, workflows, and priorities?
- Has anything become redundant or outdated?

Step 3: Validate Ownership

- Is there a clearly assigned owner for this document?
- Has that owner reviewed and approved the most recent version?
- Is the review cadence (quarterly, biannual, post-audit) documented and tracked?

Step 4: Gather Team Input

- Have we asked everyday users what feels unclear or confusing?
- Do contributors know how to suggest updates?
- Have we logged recurring points of friction?

Step 5: Archive + Refresh

- Is the previous version archived for reference?
- Has the new version been shared in the right places (project hub, Notion, intranet)?
- Do linked templates and briefs reflect the latest updates?

Processes work best when the people using them help shape them.

It's tempting to update documentation behind the scenes and roll it out with a "new and improved" label. But if your team didn't help refine it, they're less likely to use it—and more likely to revert to workarounds.

To build lasting, scalable systems, make your team co-creators of the process. They're the ones in the tools daily. They know where things get stuck, who's unclear on ownership, and what steps feel redundant. If you want your workflows to work in the real world, you need their input.

Co-Creation Reduces Friction

When your team helps shape the process, you don't have to "enforce" adoption. They've already bought in—because they helped build it.

This saves time, reduces friction, and leads to more useful, realistic documentation.

Here are some simple ways to involve your team:

- Add a standing "process feedback" item to team meetings
- Hold brief retrospectives after major content launches
- Use a shared form or Notion page to collect suggestions
- Invite rotating team members to join quarterly doc reviews

Not every suggestion will be implemented. But by showing you're listening, you create a culture where people feel invested in improving the system—not working around it.

Don't Let Feedback Get Stale

The value of team input is in the follow-up. Capture feedback in one place, set review cycles, and document decisions. When someone suggests an update, let them know what action was taken—or why it wasn't.

That loop closes the gap between observation and ownership.

I won't say "teamwork makes the dream work"—a phrase that was stenciled on the wall of an open plan office I once worked in. But it is true that, over time, creating a content program that's built on co-created processes creates a solid foundation for achieving your goals.

If it isn't documented, you can't delegate it—to a person or a machine. As your content operation grows, you'll inevitably hand off more responsibilities, whether to human collaborators or AI-powered tools. But delegation only works when your expectations, standards, and workflows are clearly documented.

Good documentation acts as the bridge between your brain and the people—or tools—helping you execute. Without it, AI outputs drift off-brand, freelancers spin their wheels, and new hires spend weeks asking the same questions.

AI Output Is Only as Good as Your Inputs

When it comes to AI, the value you get depends entirely on the quality of the inputs. Whether you're using AI for drafting, editing, or repurposing, results improve dramatically when you can provide strategic, well-structured prompts. That begins with documenting the fundamentals: your brand voice and messaging guardrails, your stylistic preferences such as formatting or headline conventions, your reader personas, your structural templates, and your definition of "done" for each type of content.

With this foundation, AI tools stop being novelties and start functioning as scalable assistants that uphold your standards. As content strategist Pamela Muldoon once told me while we were discussing podcast production, "You can't scale—or use AI effectively—if you don't even know what your process is."

Onboarding Becomes Faster—and Smarter

Clear documentation doesn't just guide machines—it accelerates human collaboration, too. New hires, freelancers, or agency partners can ramp up quickly when they have access to comprehensive resources. Instead of shadowing someone or piecing together past examples, they can self-serve style guides, voice and tone examples, step-by-step workflows, editorial calendar structures, and documented feedback and approval cycles. Even simple FAQs or "how we work" documents dramatically reduce confusion.

When onboarding is built into your documentation, new collaborators can get up to speed confidently and consistently—without draining your time or diluting your standards.

Why Documentation Unlocks Delegation

Whether you're assigning a blog post series to an AI-powered drafting tool or handing social copy to a freelancer, documentation gives everyone a fair shot at doing great work. It establishes the baseline, protects your brand, and ensures consistent quality across every channel. Most importantly, it frees you up to focus on the work only you can do—the creative, strategic, and high-value tasks that move your program forward.

I'll admit it: I wasn't always a fan of documented processes. Too often, the ones I encountered were unnecessarily complex—endless steps, too many people involved, and no clear purpose. In some organizations, every complaint about a published piece of content seemed to result in yet another process step being added. Instead of enabling quality, these steps acted as constant reminders of past failures.

Here's the truth: A process without strategy is just busywork.

As your business matures, your content operation can't remain stuck in startup mode. The scrappy systems that supported you as a one-person show—or even as a small, nimble team—won't hold up once you add clients, channels, tools, and team members. That's why your content systems must stay in active conversation with your business strategy.

When your offers change, your messaging shifts, or your audience focus evolves, your workflows, documentation, and tooling should evolve alongside them.

Workflows Should Serve Strategy—Not the Other Way Around

It's easy to cling to outdated workflows simply because "that's how we've always done it." But systems should serve your current goals, not drag yesterday's priorities into today's projects. This becomes especially critical when you're scaling up, shifting industries, or experimenting with new service lines.

The key is to pause and ask: Do our current formats still match what we're trying to achieve? Are we creating content that directly supports our offers and revenue goals? Have our review and publishing workflows adapted to the speed, compliance, or complexity our work now requires?

If the answer is no, your processes are overdue for revision.

Making Strategy Shifts Actionable

When strategy changes, it isn't enough to talk about it—you have to translate that shift into the systems your team actually uses. Update briefing templates to include new personas, buying stages, or calls to action. Adjust your editorial calendar themes so they reflect repositioned messaging. Revisit distribution checklists based on where your audience is now consuming content. And realign KPIs and reporting processes to measure against your new business goals.

Without these updates, your team may follow the documented process perfectly but still produce content that's off-strategy.

Strategy-Led Content Operations Create Room to Grow

When your systems are aligned with your goals, documentation becomes more than administration—it becomes a tool for alignment. You spend less time untangling confusion and more time creating high-impact content. And as your business grows, you're equipped to adapt without reinventing everything from scratch.

With this mindset, you're not just producing content. You're building the operational backbone of a business that's built to last.

Align Governance Updates with Business Goals and Operational Shifts

If your documentation doesn't evolve with your business, it eventually becomes a drag on progress. Processes that once felt efficient can quietly turn into obstacles as circumstances change—whether you've taken on new client types, launched a product, shifted your tech stack, or expanded your team.

That's why process updates should never happen in a vacuum. They need to be anchored in your evolving goals and operational reality. Each time the business shifts, ask yourself what that change means for how you create, review, or distribute content. Are there new stakeholders, tools, or compliance requirements that need to be documented? And, just as importantly, which old steps or roles no longer serve you?

Without this layer of strategic awareness, documentation tends to bloat with outdated processes that slow the team down or miss new expectations entirely, leaving people to guess.

Use Business Milestones as Update Triggers

Instead of waiting for a problem to surface, treat major business milestones as signals to revisit documentation. Launching a new service, product line, or campaign is a clear moment to confirm whether your systems still fit. So is hiring a new team member or changing roles internally, implementing or switching key tools like a CMS or project management platform, or resetting your KPIs and content strategy. These are the times when your systems must evolve to stay relevant—and when timely updates can prevent confusion, rework, and missed opportunities.

Make Strategic Alignment Part of Your Governance

Good documentation doesn't just support workflows—it reinforces business strategy. When you revisit your content goals, editorial pillars, or customer journey maps, your operational docs should evolve alongside them. That may mean updating persona fields in your briefing templates, refining voice and tone guidance for new markets, or shifting approval workflows to reflect new reporting lines.

When your documentation reflects not only how you work today but also where the business is heading, your team can act with clarity and confidence. Instead of being weighed down by legacy processes, they're equipped to move forward with systems that scale.

CHAPTER 9

Harnessing AI to Boost Content Efficiency

AI is changing the way content gets made. But while the tools have evolved quickly, the strategy for using them effectively hasn't always kept pace.

For content entrepreneurs and enterprise content leaders alike, the promise of AI is clear: Faster output. Smarter insights. Easier repurposing.

But without the right guardrails—clear brand voice, documented workflows, and a strong editorial process—AI can create more noise than value.

This chapter is about strategic integration. It's not about replacing human creativity or rubber-stamping AI-generated content. It's about using AI to enhance the parts of your workflow that slow you down while preserving the expertise, empathy, and brand identity that make your content work.

We'll explore:

- How to choose the right AI tools for your workflow
- Which tasks to automate vs. where human judgment matters most
- How to use AI to accelerate ideation, optimization, and repurposing
- Why governance and voice documentation are critical to scaling AI efforts

Across dozens of interviews and conversations with other market-ers, one message came through loud and clear: AI works best when it's well-guided.

Give it a clear strategy, strong inputs, and specific instructions—and it becomes a powerful partner in content creation.

Whether you're just beginning to experiment with AI or looking to operationalize it across your team, this chapter will help you build an efficient, brand-safe content engine that's ready to scale.

The Benefits of AI-Assisted Content Marketing

AI has the potential to transform content marketing from a reactive, manual-heavy process into a proactive, scalable, and insight-driven program. But to harness that potential, organizations must stop treating AI like a magic wand and start treating it like a new team member—one who requires onboarding, guidance, and structure to succeed.

Unified Brand Messaging Across Channels

When trained with your documented brand voice, content strategy, and taxonomy, AI becomes a brand amplifier rather than a wildcard. Instead of producing inconsistent tone and style across creators, AI tools help present a cohesive, on-brand presence across every channel and touchpoint.

Think of AI the way you would think of onboarding an intern: without guidance, it can't help you scale—it only produces more content that still requires your cleanup. With a brand voice chart and clear doc-umentation, however, AI strengthens consistency and ensures your message sounds unmistakably like you.

Personalized, Audience-Centric Content at Scale

With a well-defined taxonomy and robust audience personas, AI can help content creators tailor content by channel, topic, or segment. It can adapt tone and depth based on customer familiarity and even recom-mend formats aligned with audience behavior. This makes it possible to meet your audience where they are without overburdening your team.

Increased Content Production Without Sacrificing Quality

AI thrives on structure. When you provide templates, intake forms, and briefs, AI has the context it needs to generate first drafts quickly, repurpose top-performing content into new formats, and fill editorial gaps with on-brand, SEO-friendly material. Templates are not constraints—they're blueprints. They reduce cognitive load, make collaboration easier, and ensure high-quality output from both AI and human creators.

Faster, More Informed Decision-Making

Used strategically, AI supports smarter decision-making by offering real-time performance analysis, identifying high-performing assets for repurposing, and surfacing insights into audience preferences and behaviors. This data-driven feedback loop continuously refines your strategy and reduces reliance on guesswork.

Sustainable Growth and Operational Efficiency

At its best, AI is not a replacement for human marketers—it's a labor enhancer. It enables small teams to punch above their weight by automating repeatable tasks, optimizing workflows, and reducing bottlenecks.

When combined with clear governance frameworks—taxonomies, templates, guidelines, and brand voice—AI helps marketers scale operations without sacrificing strategic integrity. Governance is what stands between a cohesive content strategy and random acts of content.

AI can only deliver value when that foundation is solid.

At a Glance: How AI Delivers ROI

Benefit	How AI Supports It
Unified Brand Messaging	Trained to use and reinforce brand voice charts and guidelines
Personalized Content at Scale	Target content more precisely with comprehensive personas
Scalable Content Creation	Using templates with AI prompts delivers efficient, consistent execution
Smarter Decision-Making	Provides fast data analysis and provides concrete optimization ideas
Sustainable Productivity	Automates repeatable tasks, freeing creative bandwidth

Choose the Right AI Tools for Your Content Workflow

Not all AI tools are created equal—and not all belong in your stack. Before adopting any new technology, identify your workflow gaps.

Are you struggling with content ideation? Drafting? Optimization? Repurposing? The tools you choose should address specific needs rather than simply following trends.

Too often, teams bring in AI without a plan. They try a tool once, don't see strong results, and assume AI isn't worth the effort. In most cases, the issue isn't the tool itself—it's the lack of clarity around how to use it strategically.

Start with One Use Case at a Time

If you're just getting started, don't attempt to overhaul your entire content workflow. Pick one area where you need a boost in speed or capacity and test AI there first. For example, tools like Feedly, Perplexity, or Claude can generate timely ideas based on industry trends and audience interests. Drafting assistants such as ChatGPT, Claude, or Copy.ai can help turn raw transcripts into structured outlines or produce early drafts.

For SEO, platforms like Semrush and other AI-powered keyword generators can optimize metadata, headers, and on-page clarity. And for repurposing or formatting, Canva's Magic Write or Beautiful.ai make it easy to transform long-form content into visuals, carousels, or presentations.

The key is to start with intention. Don't test tools simply to experiment; apply them where they can meaningfully unlock efficiency without compromising your voice or quality.

Build with Your Workflow in Mind

The best AI tool is the one that fits seamlessly into your existing processes, not the one with the flashiest features. Consider whether the tool integrates with platforms you already use, if you can train or prompt it to reflect your brand voice, and if it speeds up or disrupts your team's workflows.

As content marketer and AI proponent Jason Schemmel shared in his #ContentChat conversation[12], he uses Claude and Perplexity to streamline research and idea development, but only because he has clear content structures and briefs in place. Without that foundation, the AI output wouldn't be usable.

Focus on Ongoing Use, Not Just Adoption

Simply adding AI to your martech stack doesn't guarantee impact. You still need strong inputs in the form of strategic prompts, templates, and voice guidance. You need human review to ensure nuance, context, and polish. And you need metrics to evaluate the tool's impact on both efficiency and quality.

AI tools aren't one-size-fits-all. But when used with purpose, they extend your team's capacity, reduce friction in your processes, and allow you to spend more energy on strategy and storytelling.

---

12   Schemmel, Jason. "#ContentChat Using Claude and Perplexity to Improve Content Marketing Workflows." Erika Heald Marketing Consulting, November 11, 2024. https://erikaheald.com/content-chat-how-to-use-claude-and-perplexity-for-content-marketing/. Accessed 21 October 2025.

Know When to Automate—and When Human Oversight Matters Most

AI can speed things up—but it can't make judgment calls. One of the biggest risks in AI-assisted content creation is assuming automation equals autonomy. Just because a tool can generate a blog post, draft an email, or rephrase LinkedIn copy doesn't mean it should do so without oversight.

Automation is powerful when it saves you time on repetitive, mechanical, or data-heavy tasks. But when it comes to judgment, nuance, and audience empathy, that work still belongs to humans.

What AI Handles Well

AI excels at support tasks that reduce friction without replacing strategy. It can transcribe interviews or meeting notes, summarize long-form content, generate first-draft outlines or idea clusters, suggest metadata and keywords, and reformat content from one type into another—say, turning a blog post into an email draft.

These are ideal places to experiment with automation, especially when paired with strong documentation and templates.

What Still Needs a Human Touch

By contrast, some tasks remain firmly in human hands. Establishing narrative and brand point of view requires strategy and lived context.

Fact-checking and nuance depend on subject-matter knowledge. Ensuring emotional resonance, storytelling impact, and inclusive language are decisions that no machine can make for you. And final editing for clarity, tone, and voice is where human judgment is indispensable.

AI cannot intuit your brand values or respond to subtle shifts in audience sentiment. That's why human review is not optional—it's a critical step in the workflow.

Pair Smart AI Prompts with Strong Guardrails

The more structure you give AI, the better its results. This is where your documentation does the heavy lifting. Style guides provide prompt ingredients. Templates keep structures consistent across formats. Voice and tone examples help outputs sound like you.

When your inputs are strategic, outputs require less fixing. But no matter how good the prompt, someone still has to look at the result and ask: Is this really how we want to show up?

A Practical AI Usage Decision Framework

Think of AI as a partner whose role depends on the task at hand:

- Routine or repetitive? Summarizing, reformatting, tagging, or transcribing → Use AI automation, but always review for accuracy and context.
- Creative or strategic? Storytelling, voice development, or new points of view → Human-owned tasks. AI may assist, but humans lead.
- External-facing or high-impact? Brand messaging, thought leadership, or executive communications → Always requires human review.
- Nuance required? Context-specific insights or emotional resonance → Human involvement is critical. AI can support with idea starters or tonal tweaks, but cannot own.
- Legal, ethical, or brand safety risks? Compliance-heavy or sensitive topics → Human review is mandatory.

The Bottom Line

Use AI to accelerate repeatable tasks. Use humans to elevate meaning, voice, and connection. When you combine automation with judgment, you get the best of both worlds: efficiency without losing authenticity.

How to Use AI to Accelerate Ideation, Optimization, and Repurposing

One of the most practical ways content entrepreneurs can harness AI is not to create content from scratch but to refine and remix what already exists. Used strategically, AI can help you generate smarter ideas, improve what you've already drafted, and squeeze more value out of every asset you produce.

Accelerate Ideation

AI is particularly good at breaking through blank-page paralysis. It can generate headline and angle variations for a single topic, brainstorm questions your target persona might ask, surface trending topics based on industry data, and even create draft content calendars built around evergreen themes.

That said, not every AI-generated idea deserves a spot in your workflow. It reminds me of my cat, Mister Bill: he was very good at spitting out the stale pieces of his cat food, and AI is just as good at spitting out stale, off-brand ideas. Use your judgment to sort inspiration from distraction.

Optimize Drafts More Efficiently

AI also shines as a first-pass editor. With the right inputs, it can shorten or expand content for specific channels, suggest alt text, SEO titles, and meta descriptions, identify passive voice or overly complex sentences, and highlight repetition or weak transitions. Tools like Grammarly, Hemingway, or even ChatGPT (when paired with voice and tone prompts) can make this stage faster and lighter—especially when you've already documented a comprehensive style guide.

The key is to let AI smooth the edges, not shape the voice. Your unique perspective and narrative are what make the content resonate.

Repurpose Without Reinventing

Perhaps the highest-leverage use case for AI is content repurposing. With clear instructions and structured source material, AI can turn a blog post into a video script, a LinkedIn carousel, or a newsletter intro. It can extract key takeaways for social captions, adapt content for different personas or buyer stages, and summarize assets like webinars or interviews into digestible formats.

In my own workflow, I've used tools like Canva, Copy.ai, and Beautiful. ai to transform core content pieces into multi-channel assets without reinventing the wheel each time.

A Practical Repurposing Example

Consider a blog post on "The ROI of Content Governance." With AI assistance, that single piece can quickly evolve into three social posts highlighting different points of view—one strategic, one financial, and one focused on team alignment. It can be turned into a short email recapping the key takeaway with a call to action; a slide deck or carousel that visualizes the most important statistics; an outline for a five-minute explainer video; or even a podcast intro with talking points.

All of that comes from one original asset. That's what real scale looks like—and it's powered by systems plus AI, not by burning out your creative energy.

Why Governance and Voice Documentation Are Critical to Scaling AI Efforts

Spoiler alert: AI doesn't make your brand more consistent—it makes your inconsistencies louder. When you scale content creation with AI, you don't just amplify your reach. You also amplify any gaps in your brand voice, editorial quality, or messaging clarity. That's why documentation and governance aren't optional. They are non-negotiables if you want to use AI at scale.

AI Needs Guardrails, Not Just Prompts

You can't expect to feed AI a generic instruction like "write a blog post about content strategy" and have it sound like your brand. What AI needs is structured prompts paired with brand-aligned inputs: voice guidance, persona context, and purpose. That's where governance comes in.

Governance defines the rules of the road. It provides clear standards for voice and tone, templates that shape outputs into useful formats, and style guides that keep content aligned with grammar, structure, and audience expectations. It also establishes approval workflows so that every AI draft passes through human review before it reaches your audience.

Without this infrastructure, AI becomes a content vending machine—spitting out generic outputs that may look polished but lack substance. With governance, AI becomes a creative teammate that works within your brand's boundaries.

Governance Is the Multiplier

AI can help you produce more content, faster. But only governance ensures that content is strategic, cohesive, and high-quality. When your systems are documented and enforced, you can scale without sliding into chaos. Without them, brand drift is inevitable.

Governance is the multiplier that makes AI useful. It ensures content is aligned with strategy, consistent across channels, and recognizable as yours—even when the workload multiplies.

Voice Is Your Brand's Signature

At the heart of this is brand voice. Your voice is what makes your content recognizable, relatable, and trustworthy. That's why your AI toolkit must include more than prompts—it should contain a full voice and tone guide, audience personas, prompt templates tailored to your brand, and clear examples of what good and bad content looks like in your context.

The more AI understands your voice, the better it can reinforce your brand instead of diluting it. Scaling your content operation with AI isn't just about producing more—it's about producing intentionally. That starts with documenting what great content looks and sounds like for your brand.

AI helps you go faster. Governance ensures you're headed in the right direction. Together, they create a system where speed and quality reinforce each other—rather than canceling each other out.

The Triple A Content Governance Framework

Content governance isn't just about keeping your brand safe—it's about building a sustainable foundation for consistent, strategic, and scalable content. I created the Triple A Framework—Align, Approve, Amplify— to help content leaders structure governance in a way that supports creativity, empowers teams, and drives measurable business results.

Rather than treating governance as an afterthought or a roadblock, this model reframes it as a collaborative engine that powers clarity, consistency, and reach.

Align: Ground Governance in Shared Strategy and Standards

Governance begins with clarity. The first phase, Align, is about defining what "good" looks like. This is where teams document not only voice and tone but also the purpose behind their content and the audiences they serve.

Alignment creates the strategic foundation from which every con- tributor—whether a freelancer, agency partner, or cross-functional stakeholder—can work. It ensures content reflects brand values, supports business objectives, and connects with the right audience in the right way.

In practice, this means documenting a content strategy tied to business goals, creating voice and tone guidance with before-and-after examples, mapping personas and journey stages to inform messaging, establishing content pillars and taxonomy systems that reflect audience intent, and articulating editorial principles that set your content apart from competitors.

When alignment is strong, ambiguity disappears. Creators are empowered to make decisions that reinforce strategy instead of relying on guesswork or chasing trends.

Approve: Operationalize Governance in Daily Workflows

Once the foundation is clear, governance must become part of the way work actually gets done. Approve is about operationalizing governance—turning guidelines into living systems embedded in workflows, tools, and team culture.

Governance only works when it is built into creation, not bolted on at the end. This phase integrates structure into daily processes without overwhelming contributors or slowing production.

Examples include briefing and draft templates that reflect brand standards, editorial review workflows with defined roles and expectations, onboarding resources that teach (not just tell) what's required, shared calendars or trackers that create visibility, and review checklists for voice, accessibility, formatting, and compliance.

When governance is adopted at this level, friction decreases. Teams follow repeatable processes that support creativity, focusing on meaningful work instead of rework.

Amplify: Scale What Works Across the Ecosystem

Once content is aligned and governance has been adopted, teams can move into Amplify—the stage where governance enables scale. Amplify is about reusing, repurposing, and extending the reach of what works without diluting quality or exhausting the team.

This is the difference between one-and-done publishing and strategic content operations. Amplification includes refreshing evergreen content based on performance, repurposing core assets into new formats, adopting modular content systems that can be reassembled across campaigns, enabling subject matter experts to contribute within defined guardrails, and planning distribution in sync with business events and audience rhythms.

At this stage, content leaders shift from managing production to maximizing performance. Governance makes it possible to scale impact without sacrificing consistency.

AAA Framework for Building Content Governance That Works

Phase	Focus	Key Outcomes
Align	Strategic clarity and shared standards	Brand-aligned content from all contributors
Approve	Repeatable workflows and documentation	Consistent, efficient execution
Amplify	Smart scaling and content reuse	Increased impact without extra workload

The Triple A Framework transforms governance from a reactive function into a proactive strategy. It brings everyone onto the same page, embeds best practices into daily work, and equips teams to do more with the resources they already have.

When done right, governance doesn't limit creativity—it protects it.

And, at scale, it amplifies it.

The Triple A Content Governance Model Applied to AI

The same three principles that underpin content governance also apply when integrating AI into your content program—although the details differ. This model provides a scalable framework for weaving AI into content marketing in a way that supports strategy, strengthens consistency, and maintains trust.

Rather than treating AI as a one-off experiment or a novelty tool, the Triple A Model positions it as a thoughtful component of your governance structure. It's not about controlling every output. It's about empowering your team to use AI intentionally, transparently, and in service of your content strategy.

Align: Match AI Use to Brand Purpose and Voice

The first phase is about ensuring AI usage reflects your brand's purpose, values, and voice before a single prompt is entered. Without alignment, AI tools can generate content that looks polished but misses the strategic mark.

To achieve alignment, document your brand's voice and audience intent in your style guide, create prompt templates that reflect your editorial tone and themes, and train your team not only on when to use AI but also when not to. Every AI-assisted effort should be grounded in audience-first goals, not convenience. You cannot govern what hasn't first been aligned to your values and voice.

Approve: Operationalize AI in Daily Workflows

Once alignment is established, the next step is Approve: defining how AI will actually integrate into workflows. This stage is not about replacing writers. It's about enabling smarter collaboration and more efficient production.

Start by identifying specific use cases where AI adds value, such as idea generation, outlining, or repurposing. Establish guardrails to clarify when AI use is allowed, required, or restricted. Document workflows that include human oversight checkpoints and equip contributors with onboarding resources like prompt libraries, FAQs, and quality standards.

As I often remind teams: governance isn't about stopping people from using AI. It's about ensuring the way they use it serves your goals.

Amplify: Scale Impact with AI-Enhanced Strategy

The final phase of the model is Amplify: where strategy and AI combine to extend your impact. Amplify is not about flooding the world with more content—it's about making your best content work harder.

AI can help identify repurposing opportunities across formats and channels, streamline updates and refreshes during content audits, personalize messaging at scale within brand voice boundaries, and capture learnings over time to refine prompt libraries. Amplification ensures that what works gets extended, not diluted.

Making the Model Operational

The Triple A Model isn't a theory—it's a governance foundation. It integrates with your existing content systems, from briefing template and editorial guidelines to review processes, and sets the tone for ethical, efficient, brand-aligned AI use.

AAA Framework for Content Governance With AI

Pillar	Focus	Key Outcomes
Align	Strategy + Brand Fit	AI outputs feel on-brand and intentional
Approve	Process + Enablement	Teams know how and when to use AI
Amplify	Optimization + Reach	Content scales without sacrificing quality

The Triple A Model gives your team permission to explore and experiment without losing strategic focus or brand trust. It reframes AI from a tactical shortcut into a strategic enabler—one that helps your team move faster while staying true to your voice and values.

Maximizing AI Through Governance

Integrating AI into your content workflows doesn't mean abandoning governance standards—it means updating them to support new tools and opportunities. When AI adoption is guided by intentional governance, it becomes a prolific partner to your content strategy, not a source of risk or chaos.

Here's how to bring AI into your content operations responsibly, efficiently, and with purpose:

Step 1: Evaluate AI Tools

Governance begins with thoughtful selection. Start by researching and comparing AI-powered tools through the lens of your existing martech stack. The goal isn't just to ask what a tool can do, but how it fits into your current systems, permissions, and standards.

A strong fit means the tool supports structured inputs and customizable outputs, integrates smoothly with your CMS, content calendar, or DAM, and allows for human-in-the-loop workflows with clear audit trails. Don't chase flashy features—evaluate fit, transparency, and extensibility.

Step 2: Identify Opportunities

Before deploying AI broadly, pinpoint where it can add value without disrupting your team's workflow. Begin with low-risk, high-impact areas where human oversight is easy to maintain.

AI is especially useful in early-stage tasks such as brainstorming titles, hooks, and content angles; summarizing competitor research or surfacing relevant stats; drafting metadata or alt text; and reformatting existing content for different channels. Governance ensures these tasks are clearly defined, approved, and monitored so they accelerate output rather than introduce inconsistency.

Step 3: Integrate AI Seamlessly

With tools selected and use cases identified, the next step is integration. Governance guides rollout—not just adoption—by ensuring AI is embedded in the workflow rather than bolted on.

Ask where AI fits into your current content lifecycle, which templates or briefs need to be updated to account for AI inputs, and who owns the review checkpoints for AI-generated content. Seamless integration means AI supports your processes without disrupting your brand's workflow, structure, or tone.

Step 4: Measure and Optimize

Governance doesn't end with rollout. As AI becomes part of your operations, you must monitor both efficiency and quality. Track time-to-publish and output volume alongside editorial review time, revisions needed, and alignment with voice, persona, and SEO standards.

This creates a feedback loop where AI gets better, content gets sharper, and governance keeps both aligned with business goals. With consistent evaluation, you're not just using AI—you're improving it.

Scaling Smart with AI

When AI is guided by governance, you don't just scale—you scale smart. In a fast-moving content landscape, this is how you build sustainable efficiency while protecting quality, consistency, and trust.

Quick Wins for Combining Governance and AI

Governance doesn't have to be a blocker to experimentation—it can be the very thing that enables fast, effective AI adoption. With just a few intentional practices, you can tap into AI's efficiencies while staying grounded in your brand's strategy, voice, and values.

These quick wins show how to layer AI onto your existing workflows using the guardrails you already have in place.

1. Content Ideation

AI can jumpstart the creative process when paired with performance data and editorial templates. Instead of staring at a blank page, use AI to generate working titles, potential angles, or thematic clusters—then refine those ideas through the lens of your editorial standards.

Governance in action: Feed your AI tool inputs like content pillars, personas, or top-performing examples so the suggestions are directionally correct from the start.

2. SEO Copy Generation

Structured, repeatable tasks like drafting meta titles, descriptions, and suggested tags are an ideal use case for automation. AI can produce first-pass SEO copy quickly, leaving your team free to focus on higher-value content creation.

Governance in action: Embed SEO requirements into your templates so that AI-generated metadata automatically reflects best practices and ensures on-page consistency.

3. Content Review

AI doesn't just generate content—it can also act as a brand voice co-pilot. Use it to review human-written drafts for tone, structure, and adherence to editorial standards. This reduces rework and helps catch drift earlier in the process. I also like to ask it to grade me on how well the content meets the objectives outlined in my content brief and how to improve that alignment.

Governance in action: Position AI as a coach, not a judge. Run drafts through your guidelines before they reach the editor to save time while keeping the human judgment intact.

4. Content Repurposing

Repurposing is where governance meets scale. Feed AI your longform, human-generated content and use it to produce channel-ready variations such as social snippets, summaries, teaser copy, or internal explainers.

Governance in action: Provide platform-specific formatting rules and tone guidance so the repurposed outputs feel human and remain consistent with your brand voice.

Why This Works

When governance and AI work together, you get the best of both worlds: creativity sparked faster, content produced more consistently, and a feedback loop that strengthens both your processes and your results.

These aren't distant possibilities—they're changes you can implement this week.

CHAPTER 10

Expanding Your Content Team

Build the structure that makes sustainable content growth possible. As your content efforts mature, so must your support system.

What starts as a one-person operation—or a small but scrappy team—can quickly outgrow its ability to keep pace with audience needs, publishing velocity, and strategic demands.

That's when it's time to scale your content team. But effective scaling doesn't mean hiring wildly or handing off work without context. It means building a team intentionally—using documentation, systems, and strategic roles that support both quality and growth.

Whether you're a content entrepreneur growing from solo to supported, a marketing leader expanding an internal team or outsourcing key functions, or a consultant helping clients operationalize and scale their content program, you need more than "more hands." You need the right talent, plugged into the right systems.

In this chapter, we'll explore:

- When it's time to expand your team—and what kind of support makes sense
- How to balance in-house roles with freelancers, agencies, and fractional experts
- The systems and documentation that make onboarding seamless
- How AI tools can augment your team without replacing creative expertise
- What scalable collaboration looks like across content, PR, sales, and ops

The path to scale isn't just about delegation—it's about creating structure and clarity so that everyone contributes at their highest value. With the right approach, your team becomes a growth multiplier—not a bottleneck.

Know When It's Time to Expand Your Team—and What Kind of Support You Need

Scaling your content function isn't just about adding people. It's about adding the right kind of capacity. One of the biggest challenges in content growth is knowing when to stop doing it all yourself—or when to stop asking a too-small team to carry too much weight. By the time the need for help is obvious, teams are often already underwater, and the scramble to hire or outsource creates more chaos than clarity.

The earlier you anticipate growth points, the smoother the transition will be. That starts with identifying the kind of support you truly need—and the type of work that's slowing you down.

Signs You're Ready to Expand

Regardless of your team's size, a few red flags typically indicate it's time to scale: Deadlines are consistently slipping, or quality is suffering because the pace isn't sustainable. Strategic initiatives get pushed aside for tactical execution. Your backlog of ideas grows longer even as your publishing cadence falters. Ambitious programs like thought leadership, SEO, or repurposing never seem to launch. And perhaps most importantly, either you—or your team—are showing clear signs of burnout.

When these symptoms appear, it's time to look at your workload through the lens of roles, not just tasks.

Matching the Help to the Need

The right support structure depends on your goals and your gaps. Freelancers are ideal for specialized content creation, short-term bandwidth boosts, or flexible execution. The best fit will be those with industry expertise and a strong ability to adapt to your voice.

Agencies are well-suited for multi-channel campaigns, large content volumes, or cross-functional projects. They bring breadth, but only deliver value if you have internal bandwidth to handle reviews and approvals—otherwise bottlenecks shift rather than disappear.

Fractional content leaders provide strategic guidance, mentorship, or a bridge during leadership transitions. To make this work, treat them like part of the team by providing access to goals, data, and context.

In-house hires deliver long-term consistency, deeper cross-functional collaboration, and true brand ownership. Prioritize roles that can grow with your content function—managing editors, content operations specialists, or channel leads—so you build sustainable capacity, not just headcount.

Don't Skip Your Roles Audit

Before making any hiring decisions, take time to audit the roles and responsibilities that exist today. Map out the activities happening weekly and monthly. Clarify who owns what and where duplication or ambiguity exists. Identify which skills are missing or stretched too thin.

And ask honestly what could be automated, outsourced, or delegated.

This clarity doesn't just help you scale smarter. It also strengthens your business case for new headcount, budget, or vendor investment— because you can clearly show where the pressure points are and how new resources will relieve them.

Scaling isn't about throwing bodies at the problem. It's about designing the right blend of capacity—freelance, agency, fractional, or in-house— that enables your team to focus on the highest-value work while keeping pace with the business.

How to Balance In-House Roles with Freelancers, Agencies, and Fractional Experts

The most efficient teams don't try to do everything. They know who should do what—and why. As content needs grow, handling every-thing internally often creates bottlenecks. Yet outsourcing without a clear strategy can be just as damaging, leading to inconsistent quality, mismatched tone, and fragmented collaboration.

The solution is a blended model: one where your in-house team owns the core content strategy and its execution, and external partners support your growth in the right places.

Think in Expertise Layers, Not Role Silos

Instead of organizing by rigid role definitions, think about layers of expertise.

Your in-house team should anchor the brand. They own voice, strategic priorities, cross-functional alignment, and stakeholder relationships. Typically, they are responsible for planning, approvals, and final edits.

Freelancers and contractors extend your capacity once the strategy is set. They bring subject-matter expertise or production bandwidth that allows you to scale quickly without diluting quality.

Agencies are best deployed for campaigns or high-volume workstreams that require multiple disciplines—SEO, design, video, writing—working in tandem. To get full value, however, you need internal owners who can manage the agency relationship effectively.

Fractional experts provide senior-level strategy and mentorship without the cost of full-time leadership. They are especially valuable when building a content function from the ground up or bridging transitions between phases of growth.

Ownership Matters More than Headcount

No matter the mix of contributors, ownership is what keeps the operation cohesive. Someone must be accountable for the editorial calendar and content strategy, quality control and brand alignment, internal communication and approvals, and performance tracking and reporting.

This responsibility doesn't always need to sit with a full-time employee. I know this firsthand, having served as a fractional content leader many times. But it does need to sit somewhere. Without clear ownership, even the most talented team will struggle to deliver consistently.

Questions to Guide Your Talent Mix

Before adding to your team, take time to reflect on what you truly need:

- Which roles are essential to keep in-house for consistency and control?
- Which skills are better handled by specialists we can bring in as needed?
- What support do we need right now versus six to twelve months from now?
- Do we need execution, strategy, or both?

Answering these questions early reduces turnover, accelerates onboarding, and ensures you're building a content operation that adapts as your business evolves.

The Blended Content Talent Model

Core Layer – In-House Team

- Owns brand voice and strategic priorities
- Manages cross-functional alignment and stakeholder relationships
- Handles planning, approvals, and final edits

Middle Layer – Freelancers & Agencies

- Freelancers: Add subject-matter expertise and flexible production bandwidth
- Agencies: Deliver large, multi-channel campaigns or specialized work streams
- Both require clear briefs and strong internal oversight

Outer Layer – Fractional Experts

- Provide senior-level strategy and mentorship without full-time costs
- Guide team development and operational scaling
- Step in during transitions or rapid growth phases

The Systems and Documentation That Make Onboarding Seamless

Bringing in new contributors—whether freelance, full-time, or fractional—shouldn't feel like hitting reset. If you've already built solid onboarding documentation for freelancers (as covered in Chapter 6), you're halfway there. But as your team grows, onboarding requires more than a few introductory documents.

Systematizing onboarding isn't just nice to have—it's what enables speed, consistency, and confidence at every level of your content operation.

Don't Just Hand Off Tasks—Transfer Context

The most effective onboarding doesn't stop at telling people what to do. It explains *why* you do it that way. When contributors understand your strategy, your workflows, and your voice, they're better equipped to make smart decisions without constant oversight.

Effective onboarding systems provide:

- Strategic context—the goals, audiences, and brand point of view that guide your work.
- Workflow visibility—clarity on what happens when and who owns what.
- Process fluency—a practical understanding of how tools, reviews, and approvals actually work.
- Tone and format mastery—real examples of approved, published content that set the bar for quality.

When new contributors receive both the "what" and the "why," they integrate faster and contribute with more confidence.

System Elements That Accelerate Onboarding

The key to scaling smoothly is to systematize onboarding in ways that are role-based, tool-specific, practical, and easy to access.

System	What It Covers	Why It Matters
Role-Based Onboarding Tracks	Tailored documentation by role—writer, strategist, editor, etc.	Avoids information overload and helps people be productive faster
Content Creation Playbooks	End-to-end guides for each content type	Gives new contributors confidence to plug into the process
Workflow Walkthroughs	Step-by-step view of how content moves from brief to published	Build clarity and reduces handoff friction
Tool-Specific How-Tos	Screenshots or recordings showing how your team uses your CMS, project management tools, etc.	Prevents user error and saves time during tech ramp-up
Quality & Consistency Checklist	Style guide plus editing standards, in checklist or AI prompt-ready format	Keeps every contributor aligned without micromanagement

A Note on Tool Choice

The best onboarding system is the one your team actually uses. Whether it lives in Notion, Google Drive, or a simple internal wiki, the requirements are the same: it must be accessible, searchable, and regularly updated. You don't need an expensive LMS—you need a clear, modular system that grows as your team does.

Make It a Living System

The real value of onboarding comes when it evolves alongside your team. That means assigning an owner to review and update materials quarterly, embedding links to evergreen documents directly into project briefs and templates, and asking every new hire or contractor what was missing from their onboarding experience. Each round of feedback makes the system stronger.

Earlier in this book, we covered style guides, briefs, and collaboration systems. Onboarding is where those elements converge. Together, they create the connective tissue that helps new contributors ramp up quickly, stay aligned, and keep momentum going as your content program scales.

Use AI to Augment Your Team Without Replacing Human Creativity

AI shouldn't replace your people—it should help them do their best work faster. As your content function grows, one of the smartest ways to scale without over-hiring is to pair your human team with the right AI tools. But let's be clear: AI doesn't replace strategic thinking, creativity, or brand expertise. What it can replace are the bottlenecks that slow your team down—giving everyone more space to focus on high-value work.

AI Can Support, Not Substitute

AI excels at repetitive or structured tasks. It can draft first-pass content from well-designed prompts, generate derivative assets like social posts or video scripts from longform pieces, support ideation by surfacing audience FAQs or competitor scans, and reformat content for different tones or audiences. It also handles proofreading for grammar and readability, as well as tagging, categorizing, and summarizing for content operations.

But AI cannot own what makes your content distinct. Strategic decisions about what to create—and why—remain firmly in human hands. So do brand voice, nuance, thought leadership, cross-channel storytelling, and final editorial judgment. These are the places where human expertise and empathy are irreplaceable.

Pair AI with Roles, Not Instead of Roles

The most effective approach is to integrate AI directly into team responsibilities, rather than treating it as a separate function. Writers can use AI for first drafts, outlines, and SEO formatting. Editors can lean on it for brand tone checks or to streamline long pieces. Strategists can apply it to research, message testing, or audience queries. Even project managers can benefit, using AI to track content stages, deadlines, and metadata.

This model keeps people in the driver's seat while allowing AI to handle the repetitive tasks that typically drag down efficiency.

Make AI Tools a Shared Resource

AI should feel like a shared advantage, not a privilege for a select few. Equip your team with access to core tools: Grammarly or Writer. com for editorial support, Perplexity or Claude for research, Canva or Beautiful.ai for visual assets, and ChatGPT or Copy.ai for repurposing.

To make adoption smooth, document best practices, share prompt templates, and encourage team members to swap tips on what's working. AI should feel like empowerment, not replacement.

Keep the Creative Core Human

Your AI toolkit may grow, but your brand still relies on empathy, insight, and connection—qualities only your team can bring. AI can make your team faster and more efficient, but it's your people who make your content matter.

What Scalable Collaboration Looks Like Across Content, PR, Sales, and Operations

The most impactful content teams don't operate in silos—they build bridges. As your content function grows, so do the number of internal teams that influence—or are influenced by—your work. PR needs messaging consistency. Sales needs relevant, timely assets. Operations needs clear tracking and visibility. Without a collaborative structure, content risks becoming a disconnected effort rather than a unifying force.

Scalable collaboration is built on shared language, tools, and goals across departments.

From Reactive to Integrated

Smaller content teams often find themselves stuck in reactive mode— responding to last-minute requests, scrambling to meet fire drills, and juggling one-off needs from other teams. A scalable model flips that dynamic and moves content from reactive support to proactive integration.

Instead of one-off PR requests, you co-create joint editorial calendars with PR and communications. Instead of scrambling for last-minute sales enablement, you hold quarterly alignment meetings with sales.

Instead of metrics locked in marketing dashboards, you design shared campaign briefs with clear CTAs and handoffs. And instead of operations guessing which content works, you build centralized dashboards that connect ops, sales, and content performance.

This shift reframes content as a shared resource, not a siloed service.

Systems That Enable Cross-Functional Flow

Collaboration at scale requires infrastructure. Shared editorial tools like Notion, Airtable, or Asana give visibility across functions. A single source of truth for messaging, brand voice, and campaign objectives ensures consistency. Feedback loops make it easy for sales and support teams to report back on what resonates—and what's missing. And governance frameworks define who reviews, approves, and owns each piece of content throughout its lifecycle.

When these systems are in place, collaboration becomes fluid, not forced.

Avoiding Collaboration Gridlock

Cross-functional collaboration should make you more efficient, not slow you down. The danger is turning collaboration into bottlenecks—where too many people weigh in, timelines slip, and content approval becomes a team sport.

The solution is to designate point people in each function for review, clearly document what kind of input is needed (strategy, facts, quote approval), and set deadlines for that input. Keep communication channels streamlined—no sprawling email threads or comment chaos.

Content as a Growth Connector

When done right, content becomes the connective tissue across your business. It pulls together PR's storytelling, sales' customer insights, operations' structure, and marketing's strategy into a single, shared narrative.

That's not just good for efficiency. It's how you build a content function that doesn't just scale—it amplifies your impact.

Governance as a Catalyst for Innovation, Automation, and Human-Led Growth

When most teams think about content governance, they imagine restrictions—rules, red tape, and slowdowns. But in practice, governance is what unlocks growth. It's the foundation that allows teams to scale operations, invite new collaborators, and innovate with confidence.

Rather than confining creativity, strong governance removes ambiguity, rework, and inefficiencies—creating more space for experimentation and innovation.

Human-Led Growth Starts with Structure

Growth doesn't come from endlessly adding headcount. It comes from empowering the people you already have to do better work, faster.

When a content strategy is supported by clear processes, defined standards, and structured templates, contributors no longer have to guess. They know what a strong first draft looks like. They can plug into your systems and start producing meaningful work sooner.

Governance accelerates onboarding for new hires, freelancers, and subject matter experts. It gives occasional contributors the confidence to execute without fear of "getting it wrong." And it reduces the time teams spend on editing, rewriting, and clarification. Instead of bottlenecking work with endless review cycles, governance provides guardrails that empower.

Innovation Needs Clarity, Not Chaos

When teams no longer have to reinvent workflows or correct misaligned content, they free up energy to explore new formats, tools, and ideas. Clear guidelines give your team the confidence to ideate—knowing they're building on a stable foundation.

With expectations defined, creativity doesn't just survive—it thrives. Governance eliminates second-guessing, makes it easier to test new formats without losing brand cohesion, and creates a shared vocabulary across disciplines and departments.

Innovation doesn't require reinventing the wheel. It requires knowing which wheels already work and building on them strategically.

Automation Depends on Standards

You can't automate what you haven't standardized. Governance is what makes repeatability possible—and repeatability is what makes automation effective.

From AI-assisted drafting to low-code automations and production checklists, governance provides the structure automation tools need to deliver consistent, reliable results.

With standardized metadata and taxonomy, reusable templates and checklists, documented workflows, and repeatable QA steps built into your CMS or project management tools, automation stops feeling like a shortcut. Instead, it feels like smart delegation.

Growth Without Guesswork

Scaling a content team doesn't always require exponential hiring. With strong governance, you can unlock better performance from your current team, extend your reach through trusted contributors, and automate the tasks that don't need human hands—without losing your voice, strategy, or standards.

Governance is not a barrier to growth. It's the engine that powers it.

Building Flexible Governance Systems That Support Team Growth

Governance isn't meant to be static. As your content strategy evolves—with new team members, new channels, and new technologies—your governance systems must evolve alongside it. The most sustainable content organizations build flexible governance frameworks: adaptable, resilient, and rooted in outcomes rather than rigid rules.

This section outlines five principles to help teams scale governance without stalling progress.

1. Govern Outcomes, Not Just Tactics

Effective governance doesn't micromanage how work gets done. Instead, it aligns the what and why so your team can bring their expertise to the how.

Rather than prescribing every step, focus on clarifying:

- The purpose of the content—who it's for and what it's solving.
- The quality standards that must be met.
- The strategic outcomes it should drive.

When teams understand the goals and guardrails, they are free to experiment and optimize execution while staying aligned to the bigger picture. Governance becomes a support system—not a constraint.

2. Design for Workflow Resilience

Not all content follows the same path. Some pieces move through a tightly planned editorial calendar, while others arise from executive requests, market shifts, or last-minute opportunities. A flexible governance system anticipates both.

Resilient workflows support routine and ad hoc production, provide version control and rollback options for templates and guidelines, capture key decisions in documented places (not buried in inboxes or Slack threads), and embed metadata so content remains findable and connected to larger themes.

When workflows are resilient, content doesn't fall apart when conditions change.

3. Develop Modular Content Systems

Content should be built for reuse, not reinvention. Modular systems make it possible to create core assets that can be adapted for new formats, personas, or channels without starting over. This reduces production time and strengthens consistency.

To build modularity, create assets in reusable blocks—intro copy, stat callouts, CTA blurbs. Tag assets with metadata that reflects audience, stage, and topic. Store them in centralized, searchable repositories. And use AI to repackage core messages into new formats—for example, turning a webinar into a blog post, social snippets, and a series of email teasers.

Modular systems are the foundation for scalable, adaptive content operations.

4. Build Feedback Loops Into Your Governance Process

Governance isn't a one-and-done document. It should evolve based on how your team and your audience interact with your content.

Embed checkpoints in workflows so creators, SMEs, and approvers can flag gaps or friction. Use analytics and audit findings not only to improve content but also to refine processes. Establish calendar-based or performance-based triggers to sunset outdated guidance.

These loops keep governance alive and relevant—preventing it from becoming a dusty PDF no one reads. Instead, it becomes a living tool that grows with your organization.

5. Plan for the "What Ifs"

Even the best systems can falter under pressure—especially if they weren't designed for change. Resilient governance includes scenario planning for potential disruptions:

- What happens if a key stakeholder leaves?
- How do we recover if our content calendar platform crashes?
- What if AI fundamentally changes how we create and review content next quarter?

By walking through "what if" scenarios and identifying gaps, you can create governance policies strong enough to anchor your strategy and flexible enough to evolve alongside your people, platforms, and tools.

Why Flexibility Wins

Flexible governance isn't reactive—it's proactive. It ensures your content operations stay strong, scalable, and sustainable, no matter what comes next.

Where Do We Go from Here?

You've Made It!

You've explored the frameworks, examined the building blocks, and seen how governance can transform not just your content—but your business.

Now you stand at the threshold of something powerful: the ability to scale with confidence, creativity, and clarity.

This isn't about adding more to your already full plate. It's about replacing chaos with clarity, replacing random acts of content with repeatable wins, and building a system that works for you—whether you're a team of one or leading dozens.

With your workflows mapped, your templates in place, your style guide alive and breathing, and your brand voice documented, you are now equipped to:

- Harness AI as a true creative partner, guiding it to produce on-brand, high-value work that supports your strategy.
- Extract maximum ROI from every piece of content through sustainable reuse and rapid derivative creation.
- Grow without burnout, as your processes do the heavy lifting while you focus on strategy and innovation.
- Stand out in a sea of sameness with a voice and presence that is unmistakably yours.

Key Wins

The tools are in your hands. The path is clear. The next step is to take everything you've learned and make it real—turning your governance framework into the engine that powers the next stage of your growth.

For Content Creators & Entrepreneurs

Governance gives you control over your time and protects your creative energy. It frees you from the constant scramble, allowing you to focus on the ideas and formats that bring the greatest return.

Start small. Apply your framework to a single content type. Document the process, use your templates, and distribute across channels. Watch how much smoother the work flows—and how much further each idea can travel.

Key wins you can expect: more time for creativity and less rework, clear processes for repurposing and amplifying content, a steady pipeline that supports growth, and the freedom to scale without losing your unique voice. With this foundation, you can steadily grow an audience, diversify income streams, and build a business that supports your life instead of consuming it.

For Marketing Leaders

You now have a blueprint to transform your team into a strategic, results-driven powerhouse. Governance creates consistency across every channel while empowering individuals to create confidently and autonomously.

Pilot your framework on one campaign. Map every step from request to distribution, track the results, and use those wins to expand adoption.

Key wins you can expect: reduced duplication of effort, faster turn-around times, fewer bottlenecks, smoother collaboration, and campaigns you can track and optimize with confidence. Over time, you'll build a marketing program that adapts to change without chaos and delivers measurable business impact at scale.

For Agency Owners & Consultants

With governance in your toolkit, you can offer clients more than creative execution—you can deliver a replicable, efficient content engine.

Governance helps you align faster, reduce revisions, and provide a consistent client experience.

Integrate governance into onboarding. Use intake forms, brand voice charts, and template libraries to accelerate alignment and build trust from the start.

Key wins you can expect: faster client alignment, shorter ramp-up times, clearer guidelines that keep deliverables on target, fewer revision cycles, and stronger client satisfaction. Over time, this approach cements your reputation for both creativity and operational excellence, enabling you to scale your business without scaling your stress.

The Next Stage of Growth

Governance isn't the end of your content journey—it's the beginning of a new phase. With clarity and systems in place, you can build with intention, grow without burnout, and amplify your impact.

The work ahead isn't about doing more. It's about doing what matters—with consistency, confidence, and creativity.

The most beautifully designed governance framework won't make an impact if it sits ignored in a shared folder. To create real, lasting change, governance needs visible buy-in—from leadership and from cross-functional collaborators.

Different organizations require different paths to support. What works at a lean startup might flop in a matrixed enterprise. The following four models offer proven, flexible approaches for positioning governance as a strategic enabler—not just a content team initiative.

1. The Business Case Model

Best for: Enterprise environments with formal approval processes.

- This approach frames governance as both a risk-reduction strategy and a revenue-efficiency opportunity.

To build your case, anchor the pitch in metrics leadership already cares about: brand consistency, time-to-publish, campaign ROI, or compliance gaps. Walk through a clear "before and after" scenario—highlight missed opportunities or inefficiencies caused by poor governance, and then model the time, cost, or reputational impact of resolving them.

Use procurement-friendly language like standardization, workflow optimization, and risk mitigation. This model works best when presented alongside a business operations leader or PMO stakeholder, especially in organizations where cross-functional alignment is essential.

2. The Coalition Model

Best for: SMBs and mission-driven teams.

- When centralized authority is limited—or cultural buy-in matters more than top-down direction—start by building a governance coalition.

Identify collaborators already affected by content chaos: brand, product, communications, support, even sales enablement. Invite them into a lightweight working group to co-identify pain points and co-create early solutions. Position this as a time-saver for them, not extra work.

When people help shape the solution, they're far more likely to use it—and to advocate for it across their teams.

3. The Pilot + Prove Model

Best for: Change-resistant organizations or skeptical stakeholders.

- Sometimes the fastest way to gain support is to show—not tell. This model leverages quick wins as proof points.

Select one high-volume or high-impact content type—like blog posts, customer emails, or video scripts—and apply governance selectively. Use templates, review checklists, or prompt guidance. Track improvements in speed, consistency, and quality, then package the outcomes as an internal case study to share with leadership and other teams.

Once stakeholders see governance working in practice, they'll be far more open to broader rollout.

4. The Mandate Match Model

Best for: C-suite-driven environments or major transformation initiatives.

- In moments of organizational change—scaling, integrating after M&A, or adopting new technologies—governance can be positioned as the operational glue.

Tie your governance efforts directly to active leadership priorities. For example: "We can't scale our content marketing without consistent templates and voice." Mirror the language executives are already using—whether that's AI readiness, process consolidation, or customer experience. Above all, position governance as a business enabler, not a compliance burden.

This model works best when paired with visible progress updates and regular reporting aligned to executive dashboards.

Pick the Right Model for Your Moment

Governance doesn't succeed through edicts. It earns traction through clarity, alignment, and empathy. Whether you're starting with a single template or proposing a full governance framework, one of these four models can help you secure support that sticks.

Making Governance Stick

Even the most well-crafted governance framework won't make an impact if no one uses it. Adoption is not automatic—especially in environments where "governance" is often perceived as code for red tape, bureaucracy, or extra work.

To embed governance into your culture and workflows, you must position it as an enabler, not an enforcer. The following five practices shift mindsets, create momentum, and ensure your systems don't just launch—but last.

1. Rebrand It

The word governance can sound restrictive—especially to creative teams. Instead of leading with policies and compliance, frame it as a system that makes everyone's job easier: a set of tools, templates, and clarity designed for smarter work.

Language matters. Call it a content playbook, a voice and messaging toolkit, an efficiency framework, or even a creative enablement guide. When people see governance as support rather than constraint, adoption improves dramatically.

2. Promote It

Governance won't stick if it lives in a static Notion page or a forgotten Google Doc. Promote its value—visibly and repeatedly—through internal channels. Highlight real results, such as shorter review cycles, faster time-to-publish, or fewer revision requests.

Frame improvements with context: even a 15% increase in content velocity can feel transformative when tied to business outcomes.

3. Onboard It

Treat governance as a core part of onboarding, not an afterthought. Whether it's a new marketer, a freelancer, a sales team member, or a subject matter expert, every contributor should be equipped from day one with the brand voice guide, approved content templates, and submission checklists.

By integrating governance early, you avoid bad habits and set contributors up for success.

4. Share the Wins

Governance doesn't just benefit the content team—it creates clarity and efficiency for cross-functional partners. Showcase examples where governance reduced SME review cycles, streamlined approvals, clarified expectations for internal contributors, or made outsourced work more on-brand.

When other teams feel the time savings and ease, they become your strongest advocates.

5. Celebrate It

Recognition is a powerful reinforcement tool. When governance helps a campaign launch faster, or when an AI-assisted draft lands on-brand because of strong voice guidelines, celebrate the success publicly. Spotlight the people who embraced the system and delivered results.

Celebration turns governance into a source of pride—not obligation.

From Process to Practice

Making governance stick isn't about compliance. It's about building shared understanding, reinforcing good habits, and sustaining momentum. Reframe it. Promote it. Embed it. Celebrate it.

That's how governance moves from a static framework to the fabric of daily work. From process...to practice.

Next Steps for Content Creators & Content Entrepreneurs

In today's saturated digital landscape, the content entrepreneur's greatest advantage is not simply creativity—it's the ability to create with intention, consistency, and scale. Robust documentation and streamlined processes aren't just "nice to have." They are the foundation that enables you to harness AI effectively, unlock more value from every piece of content, and create a steady pipeline of sustainable, revenue-generating work.

By establishing a clear brand voice, defining repeatable workflows, building flexible content templates, and committing to consistent governance, you give yourself the freedom to grow—without burning out or losing your unique voice. These frameworks are not bureaucracy.

They're your multiplier.

The data is clear:

- One in three creators wants to tap into new revenue streams, but 58% make less than $50K/year (Deloitte[13]).

- Top challenges include growing an audience (32.9%) and finding enough time (14.4%) (Podia[14]). The Tilt[15] echoes these struggles, with 63% struggling to grow an audience, 51% to get content found, and 46% to monetize.

- Creators aim to diversify income streams (36%), grow channels (22%), and reach new audiences (16%) (Antler[16]).

- Many wish they could work fewer hours—28.5 hours/week ideally—yet they're clocking 36.5 hours/week on average (The Tilt).

---

13 "Driving lifetime value in a content creator ecosystem." Deloitte. https://www.deloitte.com/us/en/services/consulting/articles/content-creator-eco-system-and-influencer-loyalty.html. Accessed September 29, 2025.

14 Burns, Rachel. "The top 3 challenges creators face in 2022 (Results from 900+ creator survey)." Podia. May 17, 2022. https://www.podia.com/articles/creator-survey-results/. Accessed September 29, 2025.

15 2023 Content Entrepreneur Benchmark Research. The Tilt. https://www.thetilt.com/wp-content/uploads/2023/04/R3_4.24.23_TheTilt_2023_final.pdf. Accessed September 29, 2025.

16 Forsythe, Ollie. "The 2023 Creator Economy: A new direction." Antler. March 29, 2023. https://www.antler.co/blog/2023-creator-economy/. Accessed September 29, 2025.

The opportunity is this: When you have scalable, documented systems, you can finally break the "create more, work more" cycle. You can grow your audience and income without adding endless hours to your week. You can reuse, repurpose, and reimagine content faster. And you can make AI work for you—not the other way around.

As the flood of low-quality, scattershot content continues to rise, the creators who will thrive are those who lead with clarity, consistency, and strategic reuse. With these frameworks in place, you'll not only keep your business afloat—you'll chart a course for growth that is both profitable and sustainable.

Next Steps for Content and Marketing Leaders

Scaling a marketing program in today's environment isn't about producing more content—it's about producing the right content consistently, across every channel and team. Robust governance is the bridge between a well-written strategy and measurable results. Without it, even the most talented team can end up in a swirl of duplicated efforts, missed opportunities, and random acts of content.

Workflows, templates, guidelines, and a documented brand voice aren't just operational tools. They are leadership tools. They give your team clarity, reduce approval bottlenecks, and ensure that every asset reinforces your brand story. They also make onboarding faster and allow you to leverage AI in a controlled, brand-safe way—without diluting your voice or values.

The payoff?

- Greater efficiency without burning out your people
- Consistency across campaigns, markets, and languages
- Measurable ROI from every asset in your library
- The ability to scale without sacrificing quality or brand trust

Your role as a marketing leader is to create the conditions for your team's best work. By investing in governance and scalable processes now, you're not just protecting the brand—you're building a growth engine that can adapt to changing market conditions and technologies. The leaders who succeed in the next wave of content marketing aren't those who simply "do more with less." They're the ones who document, standardize, and empower—turning governance into a competitive advantage.

Next Steps for Agency Owners & Consultants

For agencies and consultants, the real differentiator is not just delivering great creative—it's delivering repeatable, scalable success for your clients. And that's exactly what robust content governance makes possible.

When you implement frameworks for workflows, templates, guidelines, and brand voice documentation, you don't just improve your internal efficiency—you increase client satisfaction and retention. You shorten ramp-up times, minimize rounds of revisions, and create a consistent experience for every client touchpoint.

With AI in the mix, the stakes are even higher. Your clients need partners who can harness AI tools strategically—integrating them into a documented process that safeguards brand voice, ensures compliance, and still leaves room for human creativity.

The benefits compound:

- Faster onboarding for new clients and team members
- Streamlined approvals and fewer delays
- Higher perceived value through process transparency
- The ability to upsell clients on scalable content packages and repurposing services

Agencies and consultants that adopt scalable governance aren't just selling deliverables—they're selling peace of mind, long-term impact, and the confidence that their client's content investment will keep paying dividends.

In a market where competition is fierce and budgets are scrutinized, your ability to offer both creativity and operational excellence will be your strongest selling point.

Acknowledgements

This book would not have been possible without the support of **Alek Irvin**, who has partnered with me on client work, been the first audience for dozens of presentations, and served as the first reader of this manuscript—providing invaluable feedback along the way.

I'd also like to thank **Ann Handley** of MarketingProfs and **Pam Kozelka** and **Joe Pulizzi**, co-founders of the Content Marketing Institute, for their friendship and encouragement, and for building communities of exceptional marketers and content creators. Their platforms gave me the chance to step on stage, share my perspectives, and grow alongside peers I deeply respect.

Finally, I'm grateful for the **#ContentChat community**, which has been a source of inspiration every Monday at noon Pacific for the past decade. Hosting these conversations with hundreds of smart, creative content professionals has taught me more than I could have imagined—and reminded me, week after week, of the generosity and brilliance of this field.

I deeply appreciate everyone who has helped me learn, grow, and share along the way—and I look forward to what we'll build together next.

About the Author

Erika Heald is a content marketing expert, consultant, and community builder who helps B2B organizations scale their content programs without sacrificing quality or brand voice. As the founder of **Erika Heald Marketing Consulting**, she partners with technology companies, membership associations, and startups to create content strategies that drive measurable business results.

For more than two decades, Erika has worked at the intersection of content, marketing, and technology. Her career spans leadership roles in-house and as an advisor, where she has guided teams in developing brand voice frameworks, content governance models, and scalable editorial processes.

Erika is also the longtime host of **#ContentChat**, a weekly online conversation where hundreds of marketers share best practices and fresh perspectives. She is a frequent keynote speaker and workshop leader at industry events including Content Marketing World, MarketingProfs B2B Forum, and the American Marketing Association (AMA).

When she's not working with clients or championing content operations, Erika can be found experimenting in the kitchen for her gluten-free food blog, *Erika's Gluten-Free Kitchen*. She lives in Sacramento, California with her partner, three rescue cats, two bulldogs, and an ever-growing collection of sci-fi novels and dahlia tubers.

Find her online at **erikaheald.com** and join the conversation on LinkedIn with the #ContentChat community Mondays at noon Pacific.

References and Recommended Resources

Books

- Casey, Meghan. The Content Strategy Toolkit: Methods, Guidelines, and Templates for Getting Content Right. 2nd ed. Berkeley, CA: New Riders, 2023.

- Crestodina, Andy. Content Chemistry: The Illustrated Handbook for Content Marketing. 6th ed. Chicago: Orbit Media Studios, 2022.

- Didner, Pam. The Modern AI Marketer in the GPT Era: How to Get Ahead with AI and Advance Your Digital Marketing Skills. Chicago: Tilt Publishing, 2024.

- Didner, Pam. The Modern AI Marketer: Guide to Gen AI Prompts. Chicago: Tilt Publishing, 2024.

- Halvorson, Kristina, and Rach, Melissa. Content Strategy for the Web. 2nd ed. Berkeley, CA: New Riders, 2012.

- Handley, Ann. Everybody Writes: Your New and Improved Go-To Guide to Creating Ridiculously Good Content. 2nd ed. New York: Harper Business, 2022.

- Harhut, Nancy. Using Behavioral Science in Marketing: Drive Customer Action and Loyalty by Prompting Instinctive Responses. London: Kogan Page, 2022.

- Magic, Jenny and Breker, Melissa. Change Fatigue: Flip Teams From Burnout to Buy-in. Tilt Publishing. 2023.

- Penn, Christopher. AI for Marketers: An Introduction and Primer. 3rd ed. Boston: Trust Insights, 2021.
- Pulizzi, Joe. Content Inc.: How Entrepreneurs Use Content to Build Massive Audiences and Create Radically Successful Businesses. New York: McGraw-Hill, 2015.
- Pulizzi, Joe. Content Inc.: Start a Content-First Business, Build a Massive Audience and Become Radically Successful with Little to No Money. 2nd ed. New York: McGraw-Hill, 2021.
- Pulizzi, Joe. The Content Entrepreneur: Critical Strategies to Accelerate Your Success As a Content Creator. Tilt Publishing, 2024.
- Robinson, Andi. The Content Puzzle... And the Missing Piece. Bookbaby, 2022.
- Rose, Robert. Content Marketing Strategy: Harness the Power of Your Brand's Voice. London and New York: Kogan Page, 2023.

Research & Industry Reports

- Content Marketing Institute and MarketingProfs. "B2B Content Marketing Benchmarks, Budgets, and Trends: Outlook for 2025." Content Marketing Institute, October 9, 2024. Accessed October 21, 2025. https://contentmarketinginstitute.com/b2b-research/b2b-content-marketing-trends-research-2025/.
- Orbit Media Studios. "The Most Effective AI Uses for Content Marketing in 2025 [New Research]." Last modified August 14, 2025. Accessed August 30, 2025. https://www.orbitmedia.com/blog/ai-uses-for-content-marketing/.
- Orbit Media Studios. "2024 Blogging Statistics: Blogger Data Shows Trends and Lessons." Last modified 2024. https://www.orbitmedia.com/blog/blogging-statistics/.

Articles by Erika Heald

- Heald, Erika. "5 Steps for Creating a Useful Content Style Guide That Makes Your Brand Shine." MarketingProfs, August 7, 2024. https://mpb2b.marketingprofs.com/2024/08/07/5-steps-for-creating-a-useful-content-style-guide-that-makes-your-brand-shine-with-erika-heald/.

- Heald, Erika. "5 Steps To Find Your Brand Voice." Content Marketing Institute, April 6, 2022. https://contentmarketinginstitute.com/content-marketing-strategy/5-steps-to-find-your-brand-voice/.

- Heald, Erika. "Content Governance Is a Must for a Successful Content Strategy." Content Marketing Institute, August 28, 2024. https://contentmarketinginstitute.com/content-operations/content-governance-is-a-must-for-a-successful-content-strategy/.

- Heald, Erika. "From Taxonomy to Templates: Essential Building Blocks to Scale Content Creation." MarketingProfs, June 25, 2025. https://mpb2b.marketingprofs.com/2025/06/25/from-taxonomy-to-templates-essential-building-blocks-to-scale-content-creation-with-erika-heald/.

- Heald, Erika. "Harness AI To Harmonize Your Brand Voice: A Step-by-Step Guide." Content Marketing Institute, May 29, 2024. https://contentmarketinginstitute.com/ai-content-creation-tools/harness-ai-to-harmonize-your-brand-voice-a-step-by-step-guide/.

- Heald, Erika. "How to Ensure Your Flexible Brand Voice Keeps Global Consistency." Meltwater Blog, October 19, 2022. https://www.meltwater.com/en/blog/creating-a-flexible-brand-voice-that-still-maintains-global-consistency/.

- Heald, Erika. "How To Write Effective Social Media Guidelines That Protect Your Brand." Content Marketing Institute, May 8, 2024. https://contentmarketinginstitute.com/content-operations/content-governance-is-a-must-for-a-successful-content-strategy/.

Erikaheald.com Resources

- Heald, Erika. "Content Governance Is Ready for Its Slowment." Erika Heald Marketing Consulting, December 23, 2024. https://erikaheald.com/content-governance-is-ready-for-its-slowment/.

- Heald, Erika. "Empowering Content Creation Across Your Organization with AI and Governance." Erika Heald Marketing Consulting, December 9, 2024. https://erikaheald.com/empowering-content-creation-across-your-organization-with-ai-and-governance/.

- Heald, Erika. "How to Use AI for Content Repurposing (An Excerpt From The Content Entrepreneur)." Erika Heald Marketing Consulting, July 15, 2024. https://erikaheald.com/content-repurposing-using-ai/.

- Heald, Erika. "Why Content Style Guides Are an Essential Brand Asset (Free Template)." Erika Heald Marketing Consulting, August 20, 2024. https://erikaheald.com/why-companies-need-a-content-style-guide/.

- Irvin, Alek. "18 Elements of a Truly Useful Brand Content Style Guide." Erika Heald Marketing Consulting, June 2, 2022. https://erikaheald.com/18-elements-of-a-truly-useful-brand-content-style-guide/.

Featured #ContentChat Guest Episodes

- Crestodina, Andy. "#ContentChat The Value of Including Visuals in Your Content Marketing." Erika Heald Marketing Consulting, September 23, 2024. https://erikaheald.com/content-chat-the-value-of-including-visuals-in-your-content-marketing/.

- Didner, Pam. "#ContentChat Writing Great AI Prompts for Content Marketing." Erika Heald Marketing Consulting, October 21, 2024. https://erikaheald.com/content-chat-writing-great-ai-prompts-for-content-marketing/.

- DiLeo, Kate. "#ContentChat Essential Branding Elements for Successful Content Marketing." Erika Heald Marketing Consulting, April 5, 2024.

https://erikaheald.com/content-chat-branding-considerations-for-content-marketing/.

- Garrett, Michelle. "#ContentChat Aligning Owned Media and Earned Media Strategies." Erika Heald Marketing Consulting, December 9, 2024. https://erikaheald.com/content-chat-aligning-owned-media-and-earned-media-strategies/.

- Gately, Lisa. "#ContentChat Generative AI and Content Marketing." Erika Heald Marketing Consulting, January 29, 2024. https://erikaheald.com/content-chat-generative-ai-and-content-marketing/.

- Graham, Melanie. "#ContentChat How to Achieve Consistency in Your Content Branding." Erika Heald Marketing Consulting, April 17, 2023. https://erikaheald.com/april-17-2023-content-chat-recap-how-to-achieve-consistency-in-your-content-branding/.

- Harhut, Nancy. "#ContentChat Using Human Behavior and Psychology to Supercharge Your Content Marketing." Erika Heald Marketing Consulting, November 18, 2024. https://erikaheald.com/content-chat-behavioral-psychology-techniques-for-content-marketing/.

- Hill, Carmen. "#ContentChat Story Mapping for the B2B Buyer's Journey." Erika Heald Marketing Consulting, January 22, 2024. https://erikaheald.com/content-chat-recap-b2b-story-mapping/.

- Jann, Maureen. "#ContentChat An Introduction to the Most Underrated Content Marketing Format—The Quiz." Erika Heald Marketing Consulting, July 15, 2024. https://erikaheald.com/content-chat-why-quizzes-are-a-valuable-content-marketing-investment/.

- Penn, Christopher S. "#ContentChat Christopher S. Penn, Using AI for Content Marketing in 2024." Erika Heald Marketing Consulting, December 11, 2023. https://erikaheald.com/december-11-2023-content-chat-recap-using-ai-for-content-marketing-in-2024/.

- Rose, Robert. "#ContentChat Why a Fractional Content Leader is a Smart Marketing Investment" Erika Heald Marketing Consulting, April 1, 2024. https://erikaheald.com/content-chat-fractional-marketing-leaders-explained/.

- Schemmel, Jason. "#ContentChat Using Claude and Perplexity to Improve Content Marketing Workflows." Erika Heald Marketing Consulting, November 11, 2024. https://erikaheald.com/content-chat-how-to-use-claude-and-perplexity-for-content-marketing/.
- Trask, Marcia. "#ContentChat Measuring Content Marketing's Results." Erika Heald Marketing Consulting, December 2, 2024. https://erikaheald.com/content-chat-how-to-measure-content-marketing-results/.

www.ingramcontent.com/pod-product-compliance
Lightning Source LLC
Chambersburg PA
CBHW071605210326
41597CB00019B/3407